Shooting Video

Shooting Video

FREDERIC W. ROSEN

FOCAL PRESS
Boston London

Focal Press is an imprint of Butterworth Publishers

Library of Congress Cataloging in Publication Data

Rosen, Frederic W.
Shooting video.

1. Video tape recorders and recording—Amateurs' manuals. 2. Home video systems. I. Title.
TK9961.R67 1984 621.388'332 83-23128
ISBN 0-240-51709-1

10 9 8 7 6 5 4 3 2 1

Cover photograph by Don Banks.

Vidcams courtesy of Olympus and Minolta.

Butterworth Publishers
10 Tower Office Park
Woburn, MA 01801

Printed in the United States of America.

For Murray, who'll always be there,
under the clock on Seventh Avenue.

Contents

Acknowledgments

The author would like to acknowledge the following individuals, whose selflessness made this book possible: Don Banks, Mara Bovsun, Frank Capra, Ken Evans, David Fox, Irene Greenstein, Morris Greenstein, Bob Johnson, Marguerite Klein, Arlyn Powell, Anthony Regina, Addie Rosen, Ruth Rosen, George Schaub, Isidore Werner, and most of all, Ted Schwarz.

1

Toward A Video Aesthetic

Recently, a friend of mine and I were eating at a delicatessen while discussing the revolution in home video that's been going on for the past few years.

"What I want to know," my friend asked, "is why no writing is being done today about visual aesthetics in video."

Visual aesthetics? Words like that always seem a bit ambiguous to me, so I told him to be more specific.

"Why is it that the people who use video equipment and write about it aren't telling others what a really good video tape should look like? *That's* what I mean by visual aesthetics."

I smiled and told him that that was precisely what my book is about: how to actually take the video equipment — the camera, recorder, battery pack, and mike — and go out and shoot tapes using visual language.

"Even in a simple shooting situation," I explained, "where someone is going out to shoot a tape, of let's say, their kids, their wife, or their vacation, they have to know what types of shots and camera movements they're capable of creating and how to combine the shots and movements to tell a story."

My friend looked puzzled. "I don't think many people are going to shoot dramatic productions on their home video equipment."

"I'm not talking about that when I say a simple story. Far too often I come across tapes with out-of-focus shots that are improperly framed, and where every shot has a zoom thrown in. The effect on the viewer is kaleidoscopic at best and boringly amateurish at worst. Simply put, the people who shoot these tapes failed to use the proper language.

"Unless the home videographer understands this *visual* language — the one that stems from the D. W. Griffith days of silent films —

Figure 1–1 The close-up is the most intimate of all shots and uniquely suited to videography. The idea of a close-up is to get in tight on a subject's face. Visually, you are giving the viewer a little more information, so he can get to know the subject better. Filmmaker D. W. Griffith was the first one to integrate the close-up into a motion picture. It has been used extensively since, as in this close-up of Meryl Streep and Kevin Kline, taken from the film *Sophie's Choice.* As with all close-ups, this shot will have even more power when it is shown on television. (*Photo courtesy CBS/Fox Video.*)

that person will never be able to shoot a tape that is anything more than amateurish."

By this time, the waitress had come over to clear our plates, and as I poured my friend and myself another beer, I thought back to why I'd decided to write a book about home video. God knows there are plenty of books on the market now that deal with the subject. But most of them, I had decided, concentrate on product reviews, while devoting too little space to the practical applications of equipment, especially in different shooting situations. On the other hand, the video magazines use up a lot of space to review prerecorded tapes, gadgets, and other video paraphernalia. The video shooter who is buying a camera so that he can go out and shoot is left out in the cold.

"Let me get this straight," my friend said, by way of summary. "The visual language you're talking about stems from film, but there's no video aesthetic or video language per se."

"You've almost got it," I responded. "The techniques of film and video are similar, but their applications aren't. For instance, why do you find that most dramatic television shows, those shot on film or videotape that will be broadcast on television, use a preponderance of close-ups as opposed to any other shots?"

"Why," he asked.

"Because the image we play back on that box in our living room or bedroom is much more intimate than what you will see on the screen at your local movie theater. Shooting techniques have to be adapted, improved on, and created for the television medium. The close-up is the most intimate of all shots. It plays better on the small screen than the large and has more visual power — which is why you see so much of it on television."

"So," my friend said, "while photography has an established aesthetic — that is, photographers whom other photographers imitate because of their original styles — and film has filmmakers who are imitated, video doesn't have an established group of videographers who can influence the styles of home videographers?"

"Exactly."

"I fail to understand," my friend replied, "why there's no real video aesthetic. After all, the medium's been around for awhile."

"The answer to that question," I continued, "is that most of TV's history has been taken up with experimentation and perfection of the instrument itself rather than development of an aesthetic. Film

Figure 1–2 A complete portable home video system includes, from left, a vidcam, a battery charger/adapter, videocassette recorder, with matching remote unit and tuner/timer. (A second vidcam is also shown here.) These units can be bought at one time or separately. And, as long as you stay in the same format, you can buy your equipment from different manufacturers.

had all those years to become a sophisticated art form. Now it's video's turn.

"The home video enthusiast, or videographer, is in the process of creating his own aesthetic. Our job as writers and shooters is to guide him and help him make the best use of his video equipment."

WHAT'S IT ALL ABOUT?

"Can you tell me a little bit more about how one approaches the creation of such an aesthetic?"

I proceeded to give my friend, as I will now give the reader, a brief outline of the material covered in this book.

After a short synopsis on the development of TV, the introductory chapter follows with a discussion of the two major video formats,

Beta and VHS, and their similarities and differences: how the cartridges are threaded, differences in size and quality of picture, tape speeds, pin connections to the cameras, and the like. The thrust of this beginning material will be to orient the reader to video equipment on the market so that he or she can make an informed purchase — or if the purchase has already been made, to explain just exactly how the equipment works.

From there, we proceed to a discussion of camera operations: How does it work? This chapter will include descriptions of the three major viewfinders in video cameras, as well as the advantages and disadvantages of each. Also covered will be the different types of lenses and the viewing angles used for each. Under what conditions will the various lenses be most useful? Consideration will be given to the balance, weight, and overall "feel" of the video cameras on the market today.

A chapter on lighting in video will discuss the important considerations when shooting at home or on location. Different kinds of lighting the videographer will face and how to adjust the camera accord-

Figure 1–3 A video camera is just as portable as its 35mm SLR counterpart. While the image size in these two pictures is different, I think you can see that the JVC GZ-S3U vidcam is not much bigger than the Nikon F3. Granted, the JVC unit is probably the most portable one around, but its video competitors are only a little bigger, so handling a vidcam is not as difficult as you might think. In fact, with a little practice, it will feel just as comfortable in your hand as your trusty SLR.

Figure 1–4 Videotape is the recording medium in home video. The videotape comes in a cassette, like the one pictured here of a Maxell videocassette. This particular cassette is made specifically for duplicators who are preparing longer-length feature movies for playback on your VCR.

ingly will be addressed. For instance, how do you set the camera's iris manually, especially during backlit situations? We'll talk about shooting situations where available light mixes with artificial light and what efficient, relatively inexpensive lights are available to the home videographer. The last part of the chapter will get into actual indoor lighting situations, specifically, where to place the lights to illuminate the subject while separating him from the background.

The chapter on using visual language, called Shooting, represents the heart of the book. Now that the reader knows how to operate all the equipment, how does he or she actually frame the shot? How does the video operator use visual language: close-ups, long shots, medium shots, and camera movements. How can visual language be employed to tell the story of a wife's tennis lesson, a child's graduation, etc.? The text will discuss zooming, with emphasis *not* on the zoom shot, but rather on using the zoom lens for its various focal lengths. Camera moves, such as panning, tilting, and tracking will also be discussed. Finally, use of a tripod, when possible, is advocated because of the steadiness and accuracy it provides.

The chapter on video portraits covers working with the subject: how to get him or her to relax in front of the camera. We'll tell how the subject's background can be used as a reflection of the personality, as well as specific relaxation techniques that will soften up even the most reluctant subjects.

After this, there follows a discussion of the uses of the tuner/timer, along with its remote-control unit. Lists of the various features of major tuner/timers on the market will be given.

The final chapter — and one of the most important sections — contains tips from video professionals on how to make money with your video equipment. Shooting weddings, sports, and occupational video portraits for money will all be covered, as well as ways to reach your market.

But first, let's get back to how video started and the origins of TV.

Figure 2–1 These pioneers of television, Dr. Alexanderson (right) and Ray D. Kell, are shown with a mechanical scanning disc in 1927. (*Photo courtesy of RCA News.*)

2

Videocassette Recorders

Beginning in 1873, scientists had long tried to come up with a way of taking light variations, turning them into signals, and transmitting them from one point to another. The problem they encountered was in trying to develop some means of accurately scanning the signal and then transmitting it to a receiving source.

A significant breakthrough in the development of modern TV came when Vladimir Zworykin, a Russian immigrant, invented the first camera pickup tube in 1923. He called it the *iconoscope*, and it's basically that invention which can be found in nearly every video camera today. The principle behind the iconoscope is that it scans an image in descending horizontal lines. This in turn generates an electric current that changes according to the intensity of light at every point of the image being scanned. The signal is then sent over the air or through a cable to a receiving device. The device in the receiver creates a point of light whose intensity also changes in relation to the signal received. This instrument, which Zworykin also invented, is called a *kinescope*, or picture tube.

Working with RCA's David Sarnoff, Zworykin went on to perfect his invention so that it could be manufactured in a small enough size to be attractive to the general public. Several decades passed before the introduction of pickup tubes for color cameras, in 1962.

THE ADVENT OF MODERN VIDEO

Betamax

It was Sony that first introduced the home videocassette machine in 1975, and they called it the *Betamax*. While most VCRs employ the VHS format, *Betamax* has become the generic term for videocassette recorders.

Sanyo, Sears, Sony, Toshiba, and Zenith all make videocassette recorders and cameras that utilize the Beta format. Beta machines record on ½-inch videotape that is stored in a cassette with two hubs (measuring 6.1 inches by 3.8 inches by 1 inch.)

There are three recording speeds in the Beta format. On a Beta recording cassette with 729 feet of blank videotape, you can record 90 minutes when the tape speed on the VCR is set at Beta I. At Beta II, a slower speed (at which picture quality is slightly grainier), you can record as long as 3 hours. At the Beta III speed, you still get a perfectly acceptable image, although with even more grain than with the Beta II speed, and recording time is as long as 4½ hours.

The tape in the Beta cartridge is threaded in the shape of the letter U, which accounts for the fact that the Beta system is sometimes identified as the *U-loading tape system.*

The Beta system is simple to operate. As soon as the cartridge is slipped into the tape compartment of the Beta machine, the tape is placed automatically against the machine's head drum, where the record and playback heads are located. Now you are ready to begin taping.

Video Home System (VHS)

JVC put the VHS system on the consumer market in 1976 to compete against Sony. Like the Beta system, it utilizes a ½-inch magnetic videotape as the recording medium. In addition, like the Beta system, the record and playback heads are mounted on a spinning drum that rotates 1800 times per minute. This works out to thirty revolutions

Figure 2–2 Vladimir Zworykin is shown here in the 1940s with an early model of the iconoscope, which he developed as an electronic eye of the television camera. (*Photo courtesy of RCA News.*)

per second, which precisely corresponds to the number of television pictures scanned in any regular television transmission in the United States and Japan. It is there, though, that the similarities between the two formats end.

The VHS cassette is larger than its Beta competitor, measuring 7.4 inches by 4.1 inches by 1 inch. It is also capable of more recording time than the Beta cassette.

On a long-playing VHS cartridge, you can record for up to 2½ hours on the fastest tape speed, called *SP*; 5 hours and 20 minutes at the slower *LP* speed; and 8 hours on the slowest *EP* speed. There is a loss of quality as the tape speed decreases from SP to EP, but with the introduction of fine-grain tapes, picture resolution has been considerably improved at the slower speeds.

The VHS cartridge contains tape that is threaded in the shape of the letter M, which is why the VHS system is sometimes identified as the *M-loading system*.

Unlike Beta, the threading of a VHS system occurs only when the machine's record or play controls are pressed. This accounts for the small delay from the time you press the button until the machine begins to record or play back.

In addition to these surface differences, the electronic systems of the VHS and Beta recorders are fundamentally different, so the machines are *not* compatible. This means that Beta tapes cannot be played on a VHS machine, and conversely, VHS tapes cannot be played on a Beta machine.

Preference

While the Beta machine can begin to record more quickly, the VHS cartridge is capable of recording longer. Picture quality is virtually the same in both formats. It is true, though, that there are more prerecorded cassettes of movies, concerts, and self-help seminars available in the VHS format. If you have not already made your purchase, the factors that will determine which format you decide upon will be the price, quality, and comfort that the equipment affords.

It is interesting to note at this point that there is nothing new about having two different types of popular recording media competing against each other. When moving pictures began "talking" in the late 1920s, there were two systems of sound recording. The first system involved synchronizing a phonograph record with the picture on the screen. The second system, which won out and was embraced by the worldwide film community, consisted of laying an optical sound track on the film itself. The sound track was then activated by a light contained in the projector in the movie theater.

Unlike the sound movie experience, there is no *one* recognized recording medium in home video as yet. This means that the videographer has a wider field of equipment — and prices — to choose from.

PORTABLE VCRs

A portable videocassette recorder will provide a larger degree of versatility than a home deck. You can record on location and then take the same recording unit back to your home and play the tape back on your television set.

Most portables weigh in the neighborhood of 8 to 10 pounds, but some of the newer models weigh as little as 5 pounds, so it is no longer necessary to be a weightlifter to carry around a portable VCR.

Other than compactness, there are a number of features that are common to all portable VCRs, regardless of the manufacturer. This is true mostly because even though the brand names of VCRs differ from unit to unit, the majority of videocassette recorders are made by one company in Japan, Matsushita (see Figure 2–3).

Among the features that are common to all VCRs are the following:

Power Supply

Like any piece of electronic equipment, a VCR needs an electric supply to power it. There are three types of power sources you can use: the battery pack that comes with the VCR, external 12-volt DC (direct current) batteries, or household AC (alternating current) power.

The battery pack that comes with the VCR is the easiest to use. To do so, remove the battery door that you will find someplace in the rear of the VCR. Insert the battery pack and then connect its plug into a socket that you will find next to the battery compartment. Then close the battery door.

The battery that comes with the VCR may not be charged up;

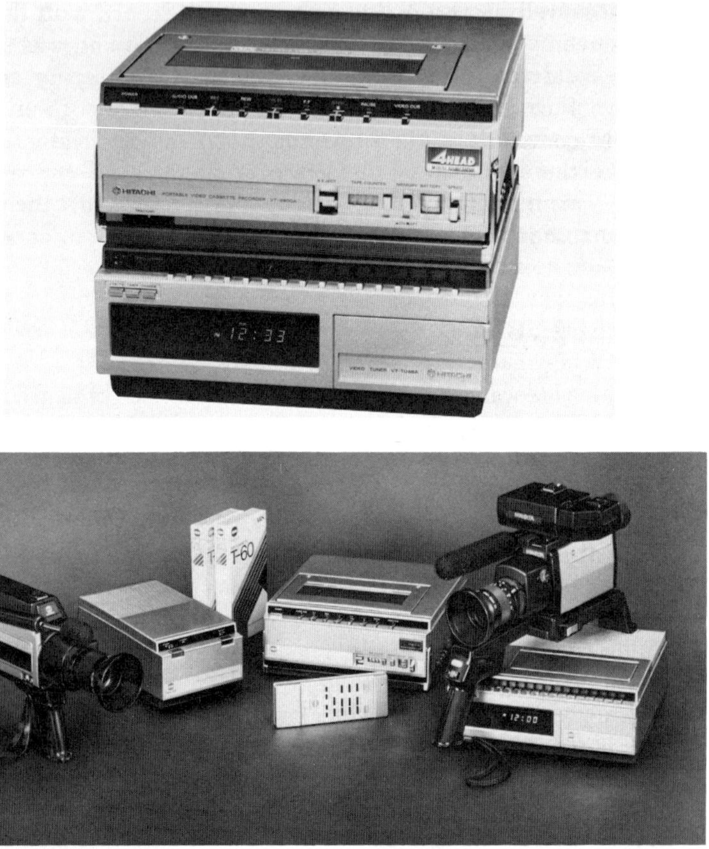

Figure 2–3 Take a good, close look at these two pictures. In the picture on the bottom, the Minolta videocassette recorder is the third unit from the right, in back of the remote control. The Hitachi VCR is on top of the tuner/timer, in the picture on the top. Notice that both VCR's are the exact, same unit. That is because they are actually made by the same company, Matsushita Electric in Japan, under contract to the respective video companies. There is a difference in price, though, which is why it is important to shop before buying.

as a matter of fact, it probably will not be. You can charge the battery pack by using either a tuner/timer or adapter/charger, either of which connects to the VCR using two or three "umbilical" cords.

The VCR can also be operated on 12-volt DC battery power. This can be accomplished by connecting the VCR to a 12-volt DC power source, such as your car's battery. The connection is made through the car's cigarette lighter, using a power cord that can be purchased from any video specialty store or from the manufacturer directly. When using these optional cords, be sure to closely follow the manufacturer's recommendations and determine first that they are compatible with your VCR. Incompatible cords or power sources could seriously damage your machine.

If you are going to power your VCR using your automobile battery, make sure that you leave the engine idling. Otherwise, the battery will discharge. Also be sure to set the VCR to the stop mode when disconnecting the cord from the cigarette lighter.

Figure 2–4 If you're trying to shoot the building of a snowman on tape, you may have problems. Your VCR's batteries will run down quicker in cold weather. Make sure to take along an extra battery pack, so you can catch all the action.

If you are in the house, you can power your VCR on simple household current. All you do is attach the VCR to your tuner/timer or adapter/charger and then plug it in. Now you can power the VCR indefinitely on household current.

In the field, the power from your in-camera battery pack is capable of giving you 20 to 30 minutes of power before recharging is necessary. That is why it is important to have an auxiliary power pack along; otherwise you may be shooting something important, only to have your power inexorably fail.

You will find that battery power will last longer in warmer weather, while in cold weather, batteries tend to drain quickly. Batteries should be kept dry and free from exposure to water and snow.

Once you are connected up to one of the three power sources just described, you are ready to begin taping.

Eject Lever

Press the eject lever located on the front of the VCR. A movable platform will spring up and remain stationary. Load your videocassette and then push the platform back into the machine until you feel it is in place.

Recording

To record, press the play and record buttons simultaneously. Make sure you have already connected your vidcam to the camera jack located on the back of the VCR.

Pausing

If you wish to pause during recording but do not wish to stop the tape completely, depress the pause button. This will stop the tape momentarily, until you press the pause button again, at which time the tape will start rolling again.

Stopping

When you have finished recording, press the stop button. This will completely stop the tape. The stop button must be pressed between recording and any other operation of the VCR. The stop button must also be pressed before you can press the eject lever to remove a cartridge.

Rewinding

To rewind what you have just shot, depress the rewind button. Press the stop button when you want to stop the tape from rewinding.

Playback

To play back what you have just shot, rewind the tape and then depress the play button. You can view the tape on the vidcam's viewfinder, if it is an electronic viewfinder, and listen to the audio by attaching an earphone to the VCR's earphone jack. Alternatively, you can look at the tape in the privacy of your home or apartment by connecting the VCR to your television set, which we will be discussed a little later on.

Fast Forward

Your VCR also comes with a fast forward button. In case you just want to view a portion of rather than all of the tape, simply depress the fast forward button until you get to the part of the tape you wish to view. Pressing the stop button disengages the fast forward mechanism.

Battery Meter

The battery meter is one of the VCR's most important features, especially when you are shooting on location. The battery meter shows the capacity of the battery pack. The green portion of the meter indicates that the battery capacity is sufficient and the red section indicates insufficient capacity. When the pointer of the meter is on red, charge the battery as soon as possible or replace it with a sufficiently charged battery pack.

Tape Speed Selector

Every VCR comes with a tape speed selector. There are three tape speeds you can choose from in each format, the advantages and disadvantages of which were outlined earlier.

Tracking Control Knob

The tracking control knob is usually located on the front of the VCR. By turning it to the left or the right, you can minimize any "noise" on the picture during playback. *Noise* is the term applied to electronic bands of interference you may see on your TV screen.

Tape Counter

The tape counter on your VCR will let you know how many feet of tape you have shot. If you depress the *counter reset button,* the tape counter will go back to zero.

CONNECTION WITH TELEVISION SET

If you do not want to view the tape you just shot through a vidcam's electronic viewfinder, you can hook the VCR up to your television set for playback. This will provide you with a color picture — in contrast to the electronic viewfinder, which can only play back in black and white, even though you have been recording in color.

RF Converter

The RF converter located on the back of the VCR changes the video and audio signals from tape playback to a standard television broadcast signal on channel 3 or channel 4 on your television set. To obtain video recorder output on either channel 3 or channel 4, set the RF converter selector switch to whichever channel is unused in your area.

Make sure that the broadcast channel on which you are viewing the videotape is not receiving *any part* of the signal from a neighboring channel. For instance, when I hooked my VCR up to my television set, I set the RF converter to channel 3, but nothing worked. It took me 2 days to figure out that the signal on channel 3, where I wanted to view my tape, was too close to the signal from channel 2. By manipulating my television set's fine tuner, I was able to tune out the signal from channel 2. Channel 3 was now filled with the snow (electronic interference) of an unoccupied channel, which I could then displace with my video signal.

Wiring

Before you can begin your playback, you still have to directly wire the VCR to the television set. Here is how to do it: Attach one end of the coaxial cable that came with the VCR to the RF out terminal

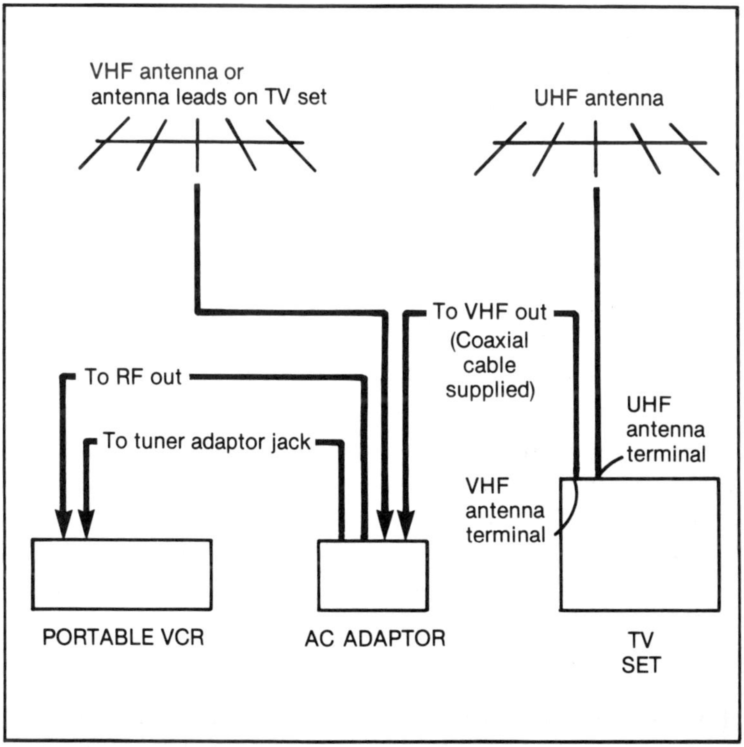

Figure 2–5 Wiring the portable VCR to the AC adaptor/charger and the TV set. (*Adapted from diagram provided courtesy of Minolta.*)

on the VCR. Attach the other end to the matching transformer that is also provided with the VCR. Then attach the transformer terminals to the VHF terminals on the back of your television set.

If you have set your RF converter switch correctly, you are now ready to play back your tape. Make sure that you double-check all your connections before beginning playback operation.

OTHER FUNCTIONS

There are a number of other functions of the VCR that you may be interested in taking advantage of:

Audio Dubbing

If you would like to add a new sound track to a tape you have already shot but want to keep the same picture, this is how you do it. First, connect the audio source, such as a tape recorder, to the audio input jack on the VCR. If you wish to record live, connect a microphone to the microphone jack of the VCR. To cue up the portion of the tape you want new sound on, play the tape back until you get to the part you want to record over. Depress the pause button at the point where you want to begin the audio dubbing. Then press the play and audio dub buttons at the same time, making sure that you are getting input from your audio source.

During the audio dub, the source that you will be recording from can be heard on your television set, so adjust the audio input level to the VCR while monitoring the TV audio. It is a good idea to adjust the TV volume to a normal setting. After dubbing, press the pause button again to release the dubbing mechanism.

If you are doing your audio dub with a microphone, turn the television's audio off or to a very low setting. This will allow you to avoid the howling type of feedback that can ruin any audio dubbing session.

Two Audio Tracks

It is also possible, if you wish, to lay another sound track over the one you recorded when you shot your tape originally, or even over a newly dubbed audio track. To do this, make your audio connections as described above, but make sure to set the sound-on-sound switch to "on." This will allow the sound on the videotape to remain, while new sounds are recorded with it. So when the tape is played back, there will be new sounds you have just added — music, for instance — together with the original sounds, such as conversation.

After using the sound-on-sound feature, be sure to turn the switch to "off." If you leave it on, the previously recorded sound is not erased but is dubbed with new sound when the next audio session takes place.

Video Dubbing

Video dubbing is especially useful for adding titles to existing sequences recorded on videotape. New video is inserted during video dubbing. However, some of the previously recorded video must be erased in the process; superimposing titles over existing footage with *video dubbing* is not possible.

The video dubbing function is also excellent for recording new shots and scenes over old ones. So if a shot turned out to be inadequate and you did not have time during the original recording session to do it again, the video dubbing function of your VCR allows you the luxury of going back and inserting the shot in its correct place on the videotape without rerecording the whole tape.

To video dub, play back the recorded cassette, turning the tracking

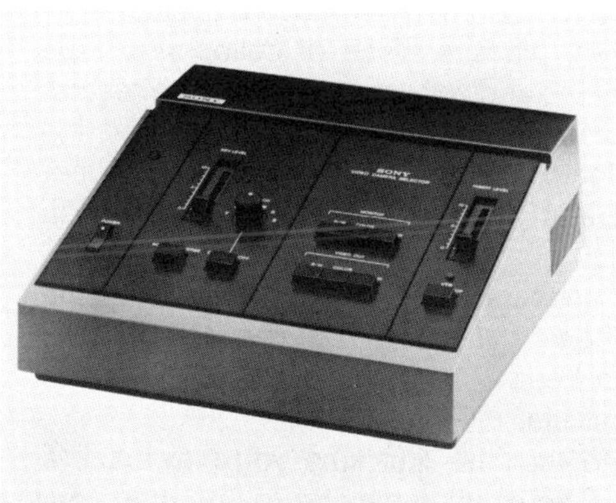

Figure 2–6 This is a special effects generator. With it, you can add some visual dazzle to your home video recordings.

knob to minimize tape noise. Find the position where video dubbing of a new shot or title is to be applied. Then, play through the shot you wish to tape over, pressing the stop button at the position desired to *stop* the video dubbing.

Press the counter reset button to set the tape counter to zero and then rewind the tape back to the video dubbing start position. Set the *counter memory* switch (an automatic stopping mechanism) to "on." Press the play button to reach the place where dubbing is to begin, and press the pause button when you have reached it. Now press the record and play buttons simultaneously, followed by the video dub button. You are now prepared to record the new scene, title, or whatever.

To begin recording, press the pause button on your VCR followed by the remote control button on your vidcam, which has been connected to the video in jack on the VCR. A new scene or title can now be recorded. The VCR will automatically stop recording at the point where the tape counter reaches zero.

PROBLEMS

Even if you follow all these directions for operation of your videocassette recorder, you might still have some problems. Here are some common ones and their simple solutions.

1. If you find that you have trouble engaging the VCR's record mode, make sure that the safety tab on the videocassette has not been removed. Some people remove it accidentally. If it has been removed, place a piece of cellophane tape over the missing safety tab and you will now be able to record.
2. If your picture is poor on playback, check to see that the television channel selector has been switched to the proper channel in your area and that the television set has been fine tuned correctly.
3. Sometimes, you will encounter video noise, the annoying streaks, globs, and snow in your picture. Frequently, this is caused by dirty video heads. The solution is to clean the heads as soon as possible.
4. Like me, you may find that the first time you play back a tape, you get nothing. The solution may be to fine tune your

television set so that your playback channel is not receiving the signal from a neighboring broadcast channel.

5. Is your playback picture somewhat noisy? You may be able to eliminate this problem by setting the tracking control to the center position.
6. If your VCR is erratic in its simple mechanical operation, check to see if your battery is fully charged. If it is not, charge or replace it immediately.
7. What happens if you find that an off-the-air television program — that is, the tape you have shot — is not being played back on your television screen? A frequent cause of this problem is a faulty connection between the VCR and the TV set. Check to make sure that all connections have been made correctly.

HOME VIDEOCASSETTE RECORDERS

With few exceptions, the controls of a home videocassette recorder, sometimes referred to as a *home deck*, are the same as the ones you will find on a portable VCR. The wiring of the home deck to the TV is also the same as the wiring between the portable VCR and the TV.

While I still think the portable VCR, coupled with the tuner/timer, is your best choice for home videography, there is a large group of people who do not wish to shoot tapes themselves, but simply want to record from their television sets. If you are in this latter group,

Figure 2–7 This is a top-loading home videocassette recorder made by Sylvania. The home deck actually contains a television tuner and videocassette recorder in one unit. Because of its weight, though, it is not recommended you use one to record in the field.

I have not left you out, as you will see when we get to the section listing VCRs.

CARE AND MAINTENANCE

If you want trouble-free operation over a long period of time, you have to care for and maintain your equipment. Otherwise, you will wind up doing something foolish, like my friend Bob. During a 16-mm shoot, Bob was assigned the task of loading a Bell and Howell camera. He did so, and then he gave the camera to his partner Jay, who did the actual filming.

Three days later, they got the processed film back and took a look at it. Every image had a foggy look, as if the sun had suddenly gone down in the middle of the shooting to be replaced by light the color of smoke.

Jay could not figure out what had happened, until Bob showed him again how he had loaded the camera. Bob had placed the base side of the film toward the lens, with the emulsion side facing away. Since it was the emulsion side that was photographically sensitive, the image was exposed through the base side, resulting in the strange, smoky-looking images.

If Bob had taken a little more time during the loading procedure, this would not have happened. While it is impossible to have something similar happen when loading your videocassette into the recorder, there are certain other things that you must watch for.

Loading

Loading is one of the places where many videographers who are just beginning make a mistake. Do not *ever* force a videocassette into a machine. If you do, you will damage the VCR's mechanism and will wind up with a costly repair bill. Also, you should not *ever* try to load a Beta cassette into a VHS machine or vice versa. It will not work.

The proper way of loading a videocassette into a recorder is, quite simply, to pay attention to the instructions on the cassette itself. At the top of the cassette, something like this will be printed: *Insert this side into recorder. Do not touch tape inside.* There is usually an arrow to show which side of the cassette gets inserted into the machine.

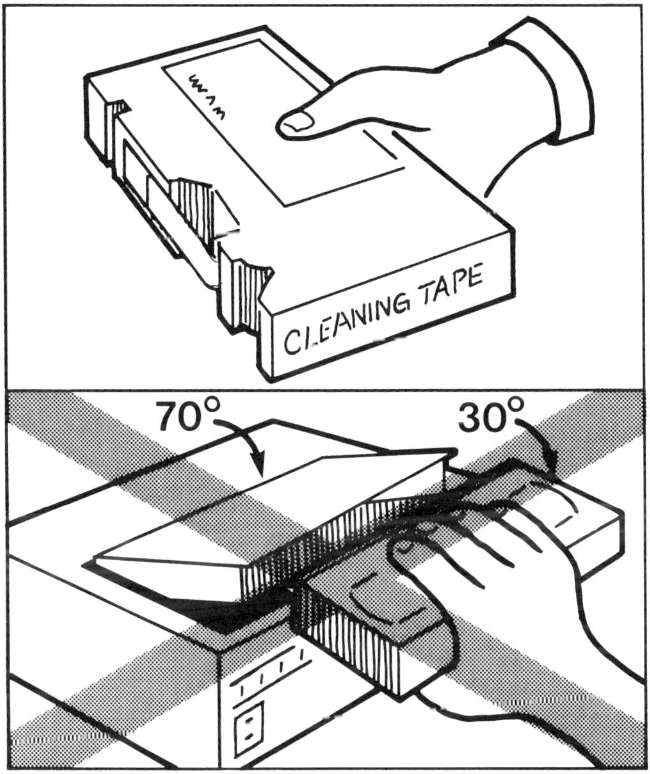

Figure 2–8 For the proper care and maintenance of your video system, one of the measures you should take is to periodically clean your machine with a cleaning tape (top). Also, you should not create damaging moisture by putting a cold tape in a warm VCR (bottom).

Videocassettes

Videocassettes are high-precision and delicate items. They require careful handling. Avoid exposing them to severe vibration, and try not to drop them.

Unlike the videotape recorder, whose tape is wound on an open spool like an audio tape, the tape used with a videocassette recorder is wound in a closed cassette. This cassette should never be opened, and the tape surface should never be touched. Doing either will not only damage the cassette, but will also damage the recording surface of the tape. You should also avoid trying to splice cassette tape. Videotape is not film; editing is done electronically, a subject we will get to a little later on.

Also remember that a videocassette cannot be turned over to play on the other side like an audio cassette can. If you try to do this, you will force the cassette into the machine and the machine will be damaged. Moreover, do not repeatedly load and unload the cassette without playing the tape. The tape inside the cartridge could become loose or even damaged.

As you will have found out during taping, the tape runs from left to right. When no tape remains on the left-hand reel, rewind it before playing it again. Otherwise, you will drain the VCR's batteries by keeping the record or play mode on while nothing is happening. The machine will *not* rewind by itself unless it has an auto rewind feature.

Temperature changes may also affect the videocassette. When a cold videocassette is used in a warm recorder, moisture may form on the tape surface, and this, in turn, will adhere to the recorder drum and cause extensive damage to the VCR. Be sure to leave the videocassette at operating temperature for a reasonable period of time before attempting to play it.

Storage

Once you have finished recording a tape or playing it back, *always* rewind the tape to the initial position. Return the videocassette to the case in which it came or a similar one purchased at a video specialty store. Store the cassette *vertically*. This means that the left-hand reel, the one with all the tape on it, will be on the bottom and the rewind reel will be on top.

In storing your videocassettes, make sure you do not put them in direct sunlight or near a heater, an oven, or any other heat source; this could damage the cassette and the tape. Look for a cool, dry place in your house — one that is also free from dust. Avoid any place that is humid as well.

Because the tape in a videocassette cartridge has a magnetic surface, you must make sure that the cassette is not stored in close proximity

Figure 2–9 Label all tapes (top) so that you can identify them and won't by mistake record over one that you want to keep. Also, don't store tapes in direct sunlight (bottom).

to a magnetic field. Avoid storing your cassette near motors, power transformers, or any household appliance that might have a magnet built into it. If the cassette should come in contact with a magnetic field, it will be erased.

There is a way to prevent accidental erasure of material you wish to retain and not record over. For instance, suppose you recorded Simon and Garfunkel's Central Park concert from your television set and you want to keep the tape to view in years to come. To make sure that you never accidentally record over this tape, break off the safety tab on the rear of the videocassette completely. Now

Figure 2–10 Have your VCR heads demagnetized every few months at a video service center (top). Another thing to avoid in the operation of your VCR is forcing connector plugs (bottom).

it is impossible to record over the cassette, even by accident, and Simon and Garfunkel will be retained for posterity.

Identification

To identify the tapes that you have shot, so you do not record accidentally over a tape that you want to keep, label your tapes. Every videocassette is packaged with labels. Simply write on the label what material a particular cartridge contains, peel it off, and stick it on either the cassette or the container you are storing the cassette in.

Maintaining the VCR

You should also obviously make sure to take care of your videocassette recorder. It is an expensive piece of equipment, and you should treat it as such.

Whenever you put it down, do so carefully; do not throw it down or drop it, even if it is in a carrying case. The carrying case will *not* cushion the machine from sudden vibrations or jolts.

When you make your connections between the VCR, the battery pack, the tuner/timer, and the camera, do not force any of the connecting plugs. If you do so, you will damage the recorder. If something does not plug in easily, then you are either using the wrong type of plug or the wrong jack for the type of plug you are using.

Maintaining the recording and playback quality of the heads of the videocassette recorder is very important. Periodically you should insert a cleaning tape into your VCR. Such tapes can be purchased at any video specialty store; when played, it will clean the tape heads of any dust or dirt. Every few months you should also take the recorder to a video service center and have the heads demagnetized. In this way, you will always be assured of crisp, sharp, and clear recordings.

PORTABLE VCRS

The following is a list of the most popular portable videocassette recorders that you may wish to consider purchasing. If you already have a VCR, you may want to compare the features of your model with those of some of the newer units. Home decks will be discussed later.

Sony

Sony is taking full advantage of the smaller Beta format videocassette. Their new Betamax portable VCR — that records, of course, in the Beta format (pioneered by Sony) — weighs only 9 pounds, 4 ounces

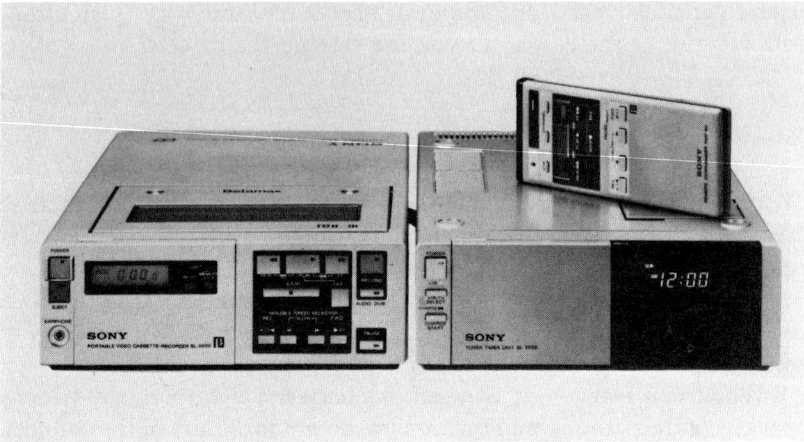

Figure 2–11 The SONY SL-2000 portable VCR is shown here with a matching tuner/timer. The tuner/timer enables you to record directly off your television set. The operation of the tuner/timer is covered in a later chapter.

and is about 30% smaller than many existing half-inch portable videocassette recorders, according to Sony.

The new Betamax SL-2000, in addition to its light weight, features a new type of picture-scanning device, called *Betascan II*, which makes the unit suitable for portable or stationary shooting, indoors or out, under a variety of shooting conditions.

Features on this unit include power on/off switch, channel up/down selection, program selection (VTR/TV), and controls for recording, play, fast forward, reverse, Betascan, stop, fast playback, pause, freeze, and Videola.

Videola, which sounds like the title of a Peter Bogdanowich movie, allows the user to view a recording in normal or slow-speed playback — either forward or in reverse. This capability makes precise picture search possible for clean, no-jump edits.

Quasar

Quasar has introduced a light, portable videocassette machine, the VP5435UQ, which records in the VHS format. This unit uses a microprocessor to assign the four heads of the machine to the appropriate recording and playback modes, resulting in greatly improved high-

Figure 2–12 This attractive young lady with too much eye make-up is holding the Quasar VP5435WQ portable VCR. This unit incorporates Dolby sound, which reduces tape hiss.

speed still and slow-motion playback with reduced electronic noise and jitter.

Another popular portable VCR that Quasar makes is the VP5420UQ. It has a stop-action picture-search control and a twelve-function, wired remote control unit.

Canon

Canon has been hard at work recently, trying to come up with a competitive portable VCR, and they think they have with their VR-10A, which records in the VHS format.

Figure 2–13 Here is Canon's portable VCR. It weighs only 8 pounds, 6 ounces. It is capable of playing back an image that you have shot at 1/6th normal speed in noiseless slow motion, or frame by frame. It can also be operated by a remote control unit, which Canon also provides.

The VR-10A weighs a scant 8 pounds, its light weight achieved through the use of integrated circuitry. It is constructed of die-cast aluminum for strength, lightness, and portability.

Its features include two tracking controls that provide both excellent playback of tapes made on other machines and elimination of noise bars when using the recorder's slow-motion feature.

Other features include electronic LCD (Liquid Crystal Digital) memory counter with built-in battery charge display (which can also be transmitted to the VC-10A camera), high-speed forward and reverse, the ability to retain memory until editing or reviewing at a later time, up to 8 hours of recording time with a long-playing tape, single-frame and noiseless slow motion, tuner input, video in/out, and audio in/out through the microphone jack, soft-touch controls, and one-touch recording. The machine is also capable of three-way operation using AC power, a 12-volt car battery, or a rechargeable battery that comes with the unit.

General Electric

Weighing in at just 8½ pounds, including battery, GE's new VHS-format, portable videocassette deck is one of the lightest on the market. GE claims that, it is 1½ pounds lighter than any comparable Beta system deck.

In addition to its light weight, this unit, whose model number is 1CVD3020X, has a full range of features, including three record/ playback speeds (SP, LP, and EP) and such special effects in the EP mode as variable slow motion, freeze frame with frame advance, and video scan up to nine times normal tape speed. There is also an elec-

Figure 2–14 One of the lightest standard format VHS portable video cassette recorders available, General Electric model 1CVD3020X (top left) can be matched to either the 1CVT605 tuner (top right) or the 1CVT615 programmable tuner (bottom left) with remote control. The deck and two tuners are shown together (bottom right).

tronic LCD digital tape counter with memory search. Moreover, the machine allows up to 8 hours of recording time.

The deck comes complete with all the necessary cables and connectors, as well as a rechargeable battery for up to 1 hour of recording between charges, GE claims. A special AC adapter that slips inside the battery compartment eliminates the need for a bulky AC power converter when using household current.

There are two other editing functions that this unit offers. First, it has an automatic tape rewind control that eliminates interference between two different taping segments. Second, there is a control that allows for insert camera recording, without interference, on a previously recorded tape.

Regardless of the external power source selected, the connection is made to the same jack on the tuner. No exotic wiring is necessary.

Minolta

The Minolta V-700S portable videocassette recorder enables videographers to tape on location with Minolta or other VHS cameras and off television sets with the Minolta tuner/timer.

The V-700S features four recording heads that provide extremely sharp images in all modes. This four-head design also permits the use of special effects in both the standard and extended play modes, a feature that is not found on all portable recorders.

Controls include sensor touch switches, and audio and video dubbing and sound-on-sound capabilities that allow easy editing. Other controls include the usual play, fast forward, reverse, pause, and stop modes.

The V-700S, which weighs 11 pounds, accepts any brand of videotape in the VHS format, providing 2, 4, 6, and 8 hours of recording time. One attractive accessory is the tape that comes with the machine. It is the T-60 Minolta-produced "how-to" tape, which gives the user a basic introduction to the uses of the Minolta recorder. This is an excellent introduction, especially for the first-time video user.

I am particularly fond of this unit, having tested it for many hours, and I could not recommend it more highly.

Zenith

It is interesting that one of Zenith's marketing slogans is "Beta is better." Although it sounds like a campaign slogan, it really shows how anxious Zenith and the other companies who make Beta products are because their VHS counterparts control more of the home video market. Now Zenith has come out with their Beta-format portable recorder.

The VR9800 is compact and light-weight. One hour of recording under the *best* recording conditions is possible on a single battery charge. The machine records in the Beta II and Beta III speeds and plays back in the Beta I, Beta II, and Beta III speeds.

Special effects are possible in the forward or reverse modes, including rapid speed search with picture, slow motion, and stop action capabilities. There is also a control that allows for clean edit control.

What I find amusing about Zenith's recorders is that they call them video "directors." When I first heard the term, I thought Zenith was making some elaborate machine that directed other machines. I was disappointed to find out that *director* was just their word for *videocassette recorder.*

JVC

Now we come to the "heavy" in the VHS class. The company that first originated the VHS format is not about to be outdone by its competitors, so it has recently introduced the HR-C3U portable videocassette recorder.

The HR-C3U is approximately 60% smaller than any previously manufactured portable videocassette recorder from JVC, weighing only 4.4 pounds (without battery pack). It allows for a maximum recording time of 20 minutes.

To add flexibility to the low-power-consumption design, power can be supplied in three ways: from nickel-cadmium battery packs, from household AC current, and from a 12-volt car battery.

Other controls on this machine, which JVC calls "conveniences," include automatic back-space editing for assembling smooth edits, a power-saving switch that ensures that energy is not wasted, a remaining time display that shows at a glance in minutes and seconds how much tape is left, two-way shuttle search, and full-function remote control.

The VHS compact videocassette used in the HR-C3U can be played on standard VHS recorders, but only by using a cassette adapter available from JVC. This could present a problem if you plan to exchange tapes with friends. You will have to remember to exchange the adapter too.

One other feature of this machine that you might find useful is that it automatically rewinds when it gets to the end of the tape.

Sanyo

One of Sanyo's latest portable videocassette recorders is the model VPR 4800. It weights a little more than some of its other Beta-format competitors — 12 pounds, including the battery.

This is a rugged recorder that comes with a nickel-cadmium battery pack that recharges completely in only 1 hour. It has a number of features worth noting.

There is the Beta-scan high-speed search system that lets you run through a tape in either forward or reverse at about seven times normal speed with the picture displayed on your television screen.

If you want to look at just one frame of the tape you have shot, there is an instant freeze-frame button that stops the picture completely. Other features of this unit include audio dubbing control, remote pulse control, digital tape counter with memory, frame advance, and a soft eject cassette door.

I should also mention that Sanyo has developed what they say is a new way of recording, the Beta III mode. Using the Beta III recording mode, you can get each Beta cassette to last 50% longer in terms of recording time, up to 5 full hours. It is a feature worth checking out.

Philco

Every time I hear the name Philco, I remember the *Philco-GoodYear Playhouse*, a fine dramatic television show that Philco co-produced in the middle 1950s. It is terribly ironic that a company that had so much to do with developing live television has now embraced the video recording medium.

In any case, Philco manufactures a full line of video equipment, including the portable V1726 videocassette recorder. It records in the VHS format.

This unit has the usual three recording speeds: SP, LP, and EP. There are controls on the recorder for video dubbing, video cue and review, and automatic stop and rewind.

Other features of the V1726 include a manual pause button, quick-charge battery, audio dubbing control, video and audio input/output jacks, and a provision for variable-speed fine slow motion.

This recorder comes with a carrying strap and a battery, and it weighs 8.4 pounds.

Sylvania

Sylvania's entry in the home video sweepstakes is the VC4526 portable VCR; it records in the VHS format.

This unit has, according to Sylvania, the capability to record up to 100 minutes with the use of *one* battery pack. I suggest you check this feature out thoroughly before buying this camera, however, since it represents so much more recording time than other machines offer.

The unit's other features include variable-speed slow motion, but only in the SP recording mode, and there is a still control for stop action viewing — again only for use in the SP mode, however. There are video and audio output jacks built into the VCR; they allow for playing TV audio through your home stereo system. There is also a Superscan control that enables you to view a taped program at nine times the normal speed, again, however, only in the SP mode.

Magnavox

Magnavox makes two portable videocassette recorders that you should be aware of. They are in the VHS format.

The VR8380 BK can be powered by the battery that comes with the unit or with household AC current. It is capable of playing back up to 8 hours of tape from a long-playing cassette. Other features

Figure 2–15 The VC-101 from Olympus has many of the capabilities of some of the larger home decks, including audio dubbing.

include three playback speeds, an electronic LCD counter, and search controls in the forward and reverse modes.

This unit is also able to give your freeze-frame effects, as well as frame-by-frame advance and slow motion. There are audio and video dub modes for editing, and when you get to the end of the tape, the machine automatically rewinds itself.

The second Magnavox portable videocassette recorder is the VR8382 BK. It has all the features of the other model, but it is capable of sixteen-function electronic remote control. Moreover, while the battery of the first unit can hold a charge for 1½ hours, this unit can hold a charge for 3 hours, according to Magnavox. I think the battery-charge capabilities of these units make them a good choice for recording in the field, if they really do hold a charge for as long as the manufacturer says they do.

Olympus

The Olympus portable VCR is durable and light. It is fully remote-controllable from the vidcam when used with a remote control unit. Its model number is VC-101, and it is a VHS unit. There are controls for slow motion and audio dub, while the single-frame advance ability of this machine allows you to view any action that you have shot in the finest detail.

SERVICE AND WARRANTIES

Before moving on to discuss video cameras, let me briefly go into something that any consumer should be concerned about: the service agreements and warranties that come with any video equipment that you buy.

Every video product is covered by a manufacturer's warranty. It is designed to give you protection in case the equipment you purchased malfunctions.

The warranties on nearly all video products are limited to a given period of time. It is usually 90 days for labor and 1 year for parts. However, warranties do differ from one company to another in terms of how long you are covered and for what you are covered. In addition, warranties differ in the methods to be used in implementing their provisions.

For instance, some manufacturers require you to pack the equipment in the original carton and ship it back to them for servicing. They may or may not pay the shipping charges one or both ways. Some, however, will supply empty cartons upon request. Some companies maintain a warranty station at the opposite end of your state which is open only a limited number of hours each day. Some require proof of purchase in order to perform free service.

Dealer Warranties

A dealer warranty may possibly extend the period of coverage on parts and labor, provided the work is done in the local service department. Here there is the convenience of being able to take your purchase back to the store where you bought it *without* the need for a shipping carton. In such a case, you would not have to pack the equipment up and ship it back to the factory or locate the nearest authorized service station yourself.

In many cases, if the unit does have to go back to the factory, dealers will provide the necessary packing and pay for the shipping themselves. Stores that do offer their own warranties usually will let you know in advance, and the warranty will be on paper if you are dealing with a reputable store.

HOME DECKS

Now let's list home videocassette recorders or home decks.

Philco

Philco offers three home VCRs. Let's start with the least expensive and move up to the top-of-the-line model. All are in the VHS format.

The V1012 has three recording speeds: SP, LP, and EP. It is capable of forward scan, frame-by-frame advance, and wired remote pause; all its controls are easily operable.

The V1441 home deck offers, in addition to the previous controls, the opportunity to freeze a picture on the television screen, electronic pushbutton VHF/UHF channel sector with an automatic fine-tuning control, and an audio dubbing button.

The V1551 is a 14-day programmable model: this is the main difference between it and its two counterparts.

Magnavox

Magnavox makes three home decks, all in the VHS format. The first unit is the VR 8316BK. It has a one-day, one-event timer, a daily

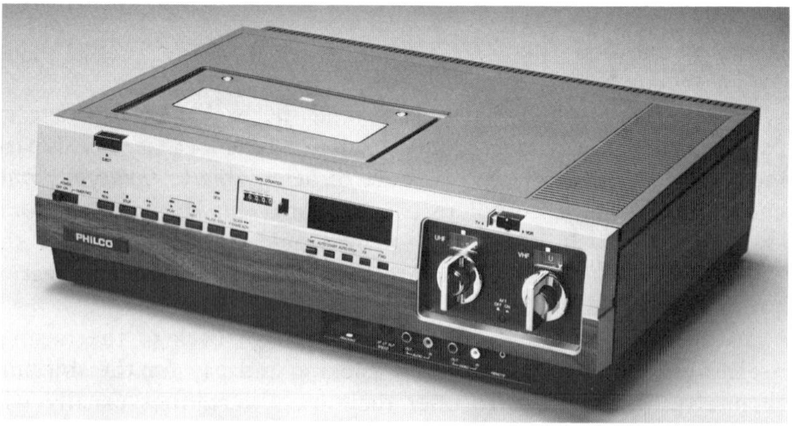

Figure 2–16 This tabletop VCR, the V1012, is the least expensive of the three home decks Philco is offering.

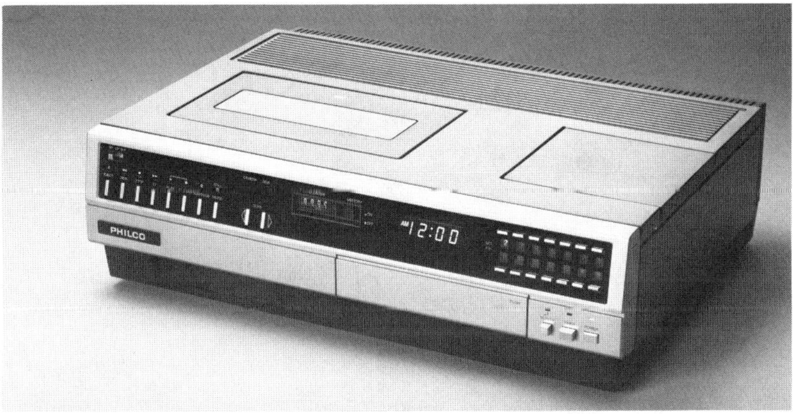

Figure 2–17 The middle-of-the line Philco VCR, the V1441. You can do audio dubbing with this unit, a function the V1012 is not capable of performing.

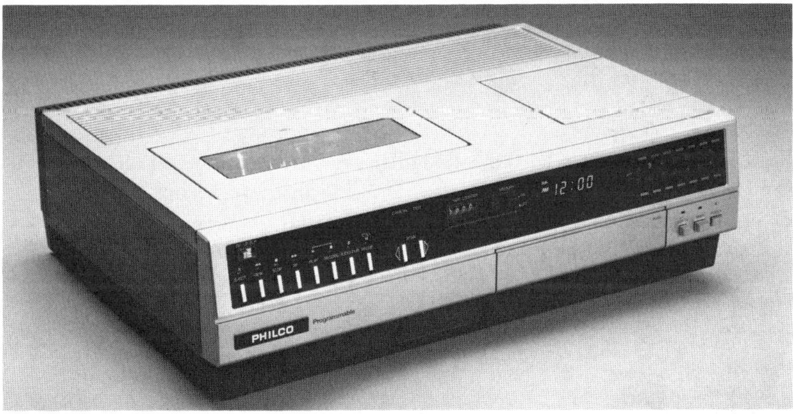

Figure 2–18 This is the top-of-the line Philco home deck, the V1551. It is a top-loading unit.

event feature, and mechanical tuning. This unit also has controls for frame advance, slow motion, auto transition editing, and remote/pause/still functions.

The more advanced VR8318BK has the same controls, with the addition of a twelve-position electronic tuner, while the top-of-the

Figure 2–19 This Magnavox portable VCR, the VR8380, is pictured with a matching charger/adapter.

Figure 2–20 This VCR, the VR8316BK, is only capable of recording one television show on one day. It is not made to record over an extended period of time, so, if you travel for days at a time and want to tape a TV show back home, this is not the unit for you.

Figure 2–21 This is the middle-of-the line Magnavox home deck, the VR8318BK.

Figure 2–22 Of the three Magnavox home decks mentioned here, this is the most expensive. Why? Because it is capable of recording eight television programs during a fourteen-day time span. This is a top-loading machine; its model number is VR8346BK.

line VR8346BK is capable of recording eight programs in a 14-day time span.

Sylvania

This company has three home decks to choose from, all VHS. The VC3620 has four record/play video heads for good performance, as well as three recording speeds. It is capable of being programmed for 14 days and eight programs.

Figure 2–23 On this Sylvania home deck, the VC2215, you have the capability of watching one program while recording another.

Figure 2–24 The Sylvania Model VC2950 is a streamlined machine. It is a front-loading unit, with controls for quickly scanning the image.

The more expensive VC2950 is a front-loading unit. It features still control for stop-action viewing, frame-by-frame advance, a digital clock, and a 24-hour timer, none of which is available on the less expensive unit.

The VC2215 has all the features of the other two units, plus a mechanical VHF/UHF tuner that allows you to watch one program while recording another.

Quasar

This company is making three tabletop VCRs; all are VHS. The VH5022UW has features that include two-way picture search, stop action, frame advance, and slow motion. There are also Synchro-touch controls, remote/pause control, and one-day, one-program programmability.

The VH5125UW is a front-loading, electronically tuned model with two-way picture search, freeze-frame capability, and the same one-day, one-program programmability.

The VH5623UW is capable of reproducing Dolby stereo. It comes with full-function wireless remote control, 105-channel tuning, automatic rewind, double-speed play, one-touch recording, and 14-day, eight-program programmability.

Hitachi

Hitachi is marketing three table-model videocassette recorders. The first unit is the VT-9710A. It has a twelve-function infrared remote and special-effects capability in the EP and SP modes, as well as forward/reverse visual search, quick, slow, still/pause, and frame-by-frame advance. The unit also has a microcomputerized timer and memory block, which are activated by a ten-key panel that allows the user to make multiple recordings and to schedule eight programs up to 3 weeks in advance.

Other features include feather-touch controls, oil-damped, soft-eject cassette, automatic playback speed selection in SP, LP, and EP modes, tone control, and auto stop and rewind at the end of the tape. There is a warning indicator in case a videocassette without the recording prevention tab is mistakenly set for timer recording, a 1-hour timer backup circuit in case of power failure, and a VTR/TV antenna switch.

The VT-15A is a four-headed machine with 2-, 4-, and 6-hour record/playback capabilities. It has 105-channel cable-ready electronic tuning, ten-function infrared remote control, visual search in the EP and SP modes with freeze frame in both modes, and a five-program, 10-day timer. The VT-15A also features auto rewind and TV channel change and includes Hitachi's Adjustomatic warranty.

The VT-11A is a two-headed machine offering 2-, 4-, and 6-hour record/playback with an electronic tuning system. It offers a one-program, 10-day timer capability with visual search in the EP and SP modes in both directions and freeze frame in the EP mode. It also has an optional eight-function remote control unit that can be purchased separately. In addition, it also carries the 2-year Adjustomatic warranty for the tape heads.

Sharp

The people from Sharp are marketing home decks, in the VHS format. Model VC3500, Sharp claims, combines the features and functions of a home VCR with the convenience and versatility of a portable unit. It weighs 15 pounds, including batteries. It can be converted to portable operation by removing the machine's powerpack and snapping in a rechargeable battery pack. The battery pack provides 120 minutes of recording time, and this unit can be used with any VHS video camera.

This unit features "one-touch recording" and a special record lock to prevent accidental recording erasure errors. There is a 7-day, one-event programmable timer that allows for unattended operation for up to a week at a time. All controls are soft-touch, and the machine allows instant and direct access to all primary recorder functions.

In the playback mode, the Model VC3500 can perform high-speed video search at five times normal speed to find particular segments of interest. Finally, each unit is supplied with remote control allowing both video search and pause/still control from across the room.

Sharp's second unit is the VC9600. It has a 105-channel compatible tuner that eliminates the need for external channel selectors and provides direct tuning of twelve VHF, seventy UHF, nine midband CATV, and fourteen superband CATV channels.

This unit has a power-assisted front-loading cassette system, automatic rewind, and one-touch recording. Other features include a 7-day, one-event programmable timer with a 12-hour AM/FM Digitron display; the timer allows the viewer to set the unit for automatic, unattended operation up to 7 days in advance. Playback highlights include high-speed video search in SP, LP, and EP modes at ten times the normal speed in both forward and reverse. There is also an optional wired remote control that allows for pause function during playback. The VC9600 also has an indicator that warns if the unit is being exposed to excessive moisture.

Figure 2–25 The Dolby noise reduction feature of this Mitsubishi table-top VCR makes it a good choice, as does its ability to tape eight programs over a fourteen-day time period. This is model number HS-320 UR.

Mitsubishi

This company has two videocassette recorders on the market. Both work in the VHS format. Model HS-320UR is a 2-, 4-, and 6-hour record/playback machine with an eight-program memory over a 14-day period. It has an eighteen-function wireless remote control that includes television power on/off and volume up/down.

Other features include Dolby™ noise reduction, elapsed time counter, seven times normal speed scan in the SP mode, nine times normal speed scan in the LP and EP modes, a counter memory, and noiseless variable-speed slow motion with still frame in the SP and EP modes.

This unit has four heads and five direct-drive motors, a frequency synchronized tuner, and is 105-channel cable ready.

Also of interest is a built-in fine-edit capability that eliminates gliches on pause and startup when recording an automatic rewind control, and a 1-hour timer backup.

The second unit that Mitsubishi makes is the HS-303UR. This unit has two heads and can record for 2, 4, or 6 hours. It has five direct-drive motors and comes with an eight-function wireless remote control.

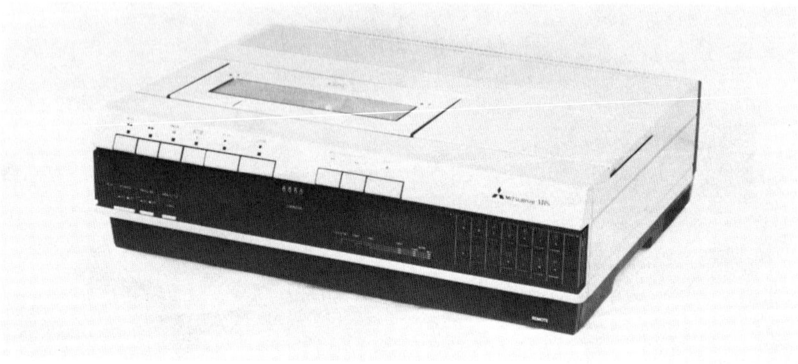

Figure 2–26 The Mitsubishi HS-303 UR is the less expensive of the two units offered, by about $500. Still, it has a fine edit capability and playback at nine times normal speed in the EP mode.

It features a fourteen-position electronic tuner, a fine-edit capability like its sister unit, nine times normal speed in the EP mode, automatic rewind, a 1-hour timer backup and a 1-day, one-event programming feature.

Akai

Akai makes a VS-7U home videocassette recorder that tapes in the VHS format. This unit has four video heads, including a slant-azimuth two-head helical scan system for better recording. With a normal T-120 VHS cassette, you can get up to a full 6 hours of recording.

All operating controls of this unit have been located on the front panel for maximum visibility and convenience. The layout has been thought out so that related controls are located close to one another for fast and accurate operation. The controls themselves are operated by a microcomputer. All loading is done through a motorized front-load system.

The unit is capable of stereo recording and Dolby™ noise reduction for superior sound quality. The VS-7U also allows you to record eight different programs over a period of up to 2 weeks. You can even designate a specific program to be recorded every day at the same time for as long as the length of the tape in your cassette allows.

Figure 2–27 The new stereo-capable videocassette recorders, such as this AKAI VS-7U, add more pleasure to the viewing experience by bringing big theater sound into the home. These VCRs can be hooked up to any high-fidelity sound system.

All settings, such as time, day, week, and channel are displayed on a large, easy-to-read, front-mounted digital display.

Recording lengths can be set in 5-minute intervals up to a full 395 minutes, and the easy-to-set "sleep timer" function shuts the recorder off automatically after a prescribed lapse of time, which can be up to 6½ hours.

The television tuner inside this home deck lets you preset up to sixteen different television channels. A synthesizer-type system has been employed for simple and accurate tuning. Adjustments can be made quickly using a single tuning control. Due to this built-in tuner, you can watch one TV program while recording another. Moreover, for those of you who have cable TV, the VS-7U can be used with the CATV system, further increasing your recording ability.

Other functions include automatic tape rewind when the tape reaches the end and a system for providing backup power for clock and timer functions, thus protecting the machine's memory for about 10 minutes in case of power failure or accidental power disconnection.

You might also be interested in the infrared remote control unit that comes with this machine. It controls such functions as power on/off, channel selection, picture search, all tape transport modes, and many more functions.

Finally, if you want to plug a video camera into the VS-7U, you can without buying a special adapter. A special camera adapter has been built into the VCR that permits direct connection to one of Akai's video cameras.

Sanyo

Beta is the recording format used in Sanyo's VCR-6300 home videocassette recorder. Sanyo claims that the VCR-6300 is the most versatile Beta-format videocassette recorder in its price range. What it has is a microprocessor-controlled tape transport system for smooth tape handling by the machine and accurate operation of all its controls.

This is a front-loading machine that allows positioning of the cartridge without the usual overhead clearance required by top-loading VCRs. At the same time, its Betascan visual picture-search control enables the viewer to scan quickly in either direction at nine times the normal tape speed. In addition, Sanyo claims that unlike other picture-search systems, their system presents a sharp and crisp image during scanning.

Other controls on this unit include instant freeze frame and a two-speed taping operation that yields up to 50% more recording time on each cassette while using the Beta III recording mode. The four-digit fluorescent tape counter on the front of the machine monitors the amount of tape used and has a memory control to help locate specific sections of the program being viewed. Moreover, the motor-assisted cassette eject mechanism handles tapes smoothly and gently without the loud "clank" associated with many VCRs.

As for programming capabilities, the VCR-6300 can be set for 7-day, single-event programming using the built-in timer. This allows you to record any show up to 1 week in advance. The large, easy-to-read fluorescent display on the front of the machine shows programming information and the time of day. The automatic fine-tuning control ensures a high-quality, high-resolution picture without additional adjustments. The electronic ten-key soft-touch tuning panel allows direct access to up to 105 channels.

Other features built into this machine include advanced noise-canceling circuitry to provide a sharper picture; extensive use in construction of LSI circuits, which results in improved performance while at the same time eliminating about two hundred components; audio and video inputs for the addition of other equipment and accessories; and audio output jacks to allow the audio portion of a program to be channeled through an outside sound system for better sound quality than your television set is capable of producing.

Sanyo also has two other less expensive home decks that should be mentioned. The VCR-4300 employs two heads for recording and playback. It features a quartz-controlled speed control that maintains a steady tape speed, the Beta-scan high-speed search system, and full remote control of all tape-transport functions, including stop, play, scan forward, reverse, rewind, fast forward, and freeze frame. There is a 7-day, single-program timer that allows recording any show you want up to a week in advance.

The third Sanyo recorder is the VCR-4200. This unit only allows timer recording up to 3 days later.

Zenith

Zenith has no less than four home videocassette recorders available in the Beta format. Model VR8500PR is a fourteen-position electronic channel selection model. It has a 24-hour timer and allows for recording in the economical Beta III mode. The unit has speed search with picture in forward and reverse and other controls for pausing to edit during recording or viewing still pictures during playback. A digital clock, tape counter, and optional wired remote control unit round out the features of the VR8500PR.

Model VR8900W has the same features as the previous model, but includes closed-caption recording capability for the hard of hearing.

Model VR9760W can be programmed for up to 14 days in advance. The unit has a clear, variable-speed slow-motion control, as well as frame-by-frame advance and stop action. The 2-week, four-event programmer has a daily or weekly repeat capability. The fourteen-position electronic channel selection mechanism operates with the automatic rewind control. Audio dubbing control rounds out this unit's features.

Model VR9775PT is the top-of-the line Beta from Zenith. It has three video playback heads instead of the usual two and is a front-loading machine.

The unit has a wireless infrared remote video action control, and it features power on/off, speed search with the picture at fifteen times the recorded tape speed, clear, multispeed slow motion, frame-by-frame advance, double speed, and stop action.

The microprocessor-controlled fourteen-position electronic channel selector enables you to change channels instantly. Other features include electronic index with cue maker, tape counter, digital clock with power-reserve function, clean edit control, and a direct camera connector.

JVC

JVC has three VHS home decks that form the core of their video equipment line. The HR-7100U is the least sophisticated and most inexpensive of the three. It is especially designed for the person who is just getting interested in home video and wants to use a good, solid unit. Even so, it is quite sophisticated.

The HR-7100U has four heads for recording and playback. It records in the SP, EP, and LP modes and the tape is driven through the cartridge and the machine by a four-motor system that includes a direct-drive motor for the head drum.

This machine has the capability of being programmed to record any program within a 10-day period or any regular series of programs. Check the setting on the fluorescent display on the front of the machine for preset taping times.

Each of the controls on JVC machines is color coded — the play and stop buttons are blue, the fast forward, rewind, and eject buttons are gray, the record button is red, and the pause button is yellow.

JVC emphasizes the visual-search capability of this machine, but the company has invented a new name for it: *shuttle search.* When you engage the shuttle-search control of this VCR, you can locate any picture that you want to view on a given tape at high speed and in either direction. The function works in the SP mode at about seven times normal speed or the EP mode at about 21 times normal speed.

There is a remote control unit that comes with the VCR, but it has to be wired into a jack on the front of the machine. Other features include a channel-lock system for preventing accidental changing of channels during recording, a tape-protection circuit that shuts the power off after 5 minutes in the pause mode, a condensation-prevention circuit, automatic rewind at the end of the tape, and a built-in RF converter for playback on any ordinary television set. This machine weighs a hefty 20.1 pounds, which explains why it is extremely difficult to take a home deck out into the field to record.

The second home deck from JVC is the Model HR-7300U. This is a top-loading machine with the same type of four-head direct-drive system as the previously mentioned VCR. The controls on this machine are different, though. They are not color-coded, although they have been strategically placed and are self-illuminating. Microprocessor-based electronic logic allows direct changeover from one mode to another.

The machine comes with a 14-day, eight-event microprocessor-controlled timer. This sophisticated timer provides much flexibility, working as if it were in fact eight independent timers; eight programs on any channel can be recorded in the 14 days after setting the controls. The HR-7300U will switch on, record one program, switch off at the end, then switch on again to record another program. Each of the eight independent memories can be used to store information related to a program that is broadcast at the same time every day or every week.

The HR-7300U also has shuttle search, backward as well as forward. In the standard play mode, the tape speed during the shuttle search is about seven times normal, while in the extended-play mode it speeds up to about twenty-one times normal.

Other features include full-function wired remote control and auto rewind, which winds the tape back to the beginning when the end of the tape is reached.

The third JVC home recorder is the HR-7650U. It offers front-panel cassette loading. Once a cassette is inserted into the front panel slot, it is automatically and quietly positioned for operation.

This unit provides an interesting editing function. When assembling a single tape from more than one segment, the automatic editing control backspaces the tape at each pause in preparation for the next segment. The result is a tape that is free of distracting, uneven edits. You can also insert material from a camera, a television, or another recorder onto an already recorded tape with minimal picture distortion at the starting and stopping points of the tape.

The HR-7650U also has stereophonic sound capability. You can combine your stereo system with this unit to create stereophonic audiovisual entertainment because of the two separate audio channels the machine provides.

Where other JVC units have a remote control that operates only with a wired connection, this top-of-the line model has a wireless infrared remote control. It controls fifteen functions of the VCR, including power on/off, channel selection, and shuttle search. You can do virtually anything but load the cassette from the comfort of your viewing position.

The programmable timer is capable of 14-day, eight-event programming. You can preset this VCR for any eight broadcasts in a 2-week period. The data in each of the eight memory cells may be checked on the fluorescent display located on the front of the machine. The quartz crystal clock will accurately and reliably carry out any of your programming commands.

Other features of this machine include a sleep timer facility, two-step tape end warning, comprehensive dimmer display with dimmer control, front-panel direct camera connection, automatic tape end rewind, 10-minute memory backup for the timer, cue and counter search, simple channel preset system, condensation-prevention circuit, easily accessible front panel V-hold control, automatic playback switching between the 2- , 4- , and 6-hour recorded tapes, and a four-digit fluorescent display tape counter.

General Electric

General Electric has four home decks that provide a variety of features for the home video enthusiast. All are in the VHS format. Let's start at the bottom of the line and work our way up.

Model 1VCR3000W gives you the option of two recording speeds for up to 8 hours of uninterrupted recording or playback on a single cassette. A brightly lit digital clock timer not only gives you the

Figure 2–28 GE's lineup of table model VHS video cassette recorders for 1983 includes model 1VCR3000W (top left), model 1VCR3010X (top right), model 1VCR3014W (bottom left), and model 1VCR3018W.

time of day, but also lets you make unattended recordings when you are away from home, busy, or asleep in the next room.

Simply set the timer, up to 10 days in advance, and it starts and stops your recording automatically for one preselected recording period of up to 8 continuous hours. The timer can also be set to record at the same time every day for as long as tape remains on the cassette.

There is a built-in pushbutton electronic VHF/UHF tuner that allows you to record a program on one channel while you watch one on another channel. You can also record a show while your TV is off. Any combination of VHF and UHF channels up to a total of twelve can be programmed into the tuner.

Tracking control, conveniently located up front, can be used during playback if stripes or noise appear in the picture; this sometimes occurs in a recording made on another VCR. A simple turn of this control, left or right, usually clears up the picture interference. Connections for pause remote control and audio and video accessories are also conveniently located in the front of the unit.

There is a digital electronic tape counter and memory lock on the machine. When in the "on" position, the memory will stop the tape during rewind when the tape counter reaches 9999.

The machine weighs a total of 21 pounds, with an average power consumption of 34 watts at 120 volts of household current.

The second GE home videocassette recorder is Model 1VCR3014W. This unit provides 8 hours of uninterrupted recording or playback on a single cassette. However, unlike the previous unit, it has five-event, 7-day programmable capabilities.

This unit also comes with a wireless multifunction remote control device. It allows you to command the recorder to perform sixteen functions: play, record, stop, rewind, fast forward, video scan (forward and reverse), quick, variable slow motion, frame advance, pause/still, slow motion, tracking, and channel change. Even if your television set is not equipped with remote control, the channel-change feature of the remote unit allows you to change TV channels through your VCR.

Up to sixteen channels can be programmed into the built-in electronic tuner. In addition to normal VHF and UHF channels, unscrambled midband and superband cable channels can also be programmed in any combination, eliminating the need for a special converter box. Make sure, though, to check your local cable company, because certain transmissions are intentionally scrambled.

There is a special program search control as well. This feature allows you to place a special electronic signal at the beginning of each new recording. This will signal the VCR to stop automatically at each separate recording segment on the tape during either fast forward or rewind, making it easy to locate the desired programming.

There are special-effects features available, but only in the EP recording mode. These include variable slow motion, quick motion, freeze frame, and freeze advance, all of which can be operated by the hand-held remote control unit.

The machine weighs 23 pounds and uses 47 watts of power from 120-volt household current.

The third GE VCR is the 1VCR3010X, which is capable of three-event, 10-day programming. A brightly lit digital clock not only gives you the time of day, but also lets you make unattended recordings. Just set the timer, up to 10 days in advance, for as many as three different times and channels. The timer will start and stop the recording automatically.

This unit also comes with pushbutton electronic tuning. The built-in twelve-channel electronic tuner allows you to record one program while you view another. Any combination of VHF and UHF channels, up to a total of twelve, can be programmed into the tuner. The unit also comes with the same features as the two preceding ones, including an audio dubbing feature and pause control. The 1VCR3010X weighs 21 pounds and consumes 34 watts of power from 120-volt household current.

The last GE home recorder we are going to consider is the most sophisticated one. The 1VCR3018W allows for eight-event, 3-week programmability. In addition to the features of the other three GE units, this one is capable of stereophonic sound.

This recorder can record from your television set and FM receiver to provide stereo simulcast. It can also dub a second sound track for sound-on-sound recording. Then it can play it all back in stereo through any stereo television or audio system. A special noise-reduction circuit operates during playback for cleaner sound reproduction, and you can also play back prerecorded VHS stereo cassettes.

The unit weighs 23 pounds, and its power consumption is 47 watts from 120-volt household current.

3

Video Cameras

Once you have decided on the videocassette recorder you are going to be using, the next step is selecting the video camera to go with it. When I was in college, video cameras were bulky things. They weighed 10 to 15 pounds and never felt very comfortable in your hands. Today's new breed of video camera is light, versatile, and less expensive than the earlier models. In addition, the manufacturers, be they camera companies that have expanded into the video line or video companies that have always specialized in video equipment, never hesitate to call in internationally famous designers to create stylish lines for their new vidcams.

There are a number of general and specific selection criteria you should use when considering a vidcam. Let's deal with the general items first.

COMFORT

Recently, I was pricing video cameras in a famous electronics store at my local shopping mall. A bright young man fresh out of high school was the salesman. He answered all my questions about a video camera that he was displaying on a huge, impressive-looking tripod. However, when it came time to try it out, he would not allow me to do so.

Upon further questioning, he said that he had had quite a bit of trouble screwing the camera on to the tripod, so he never took it off. I complimented him on his ability to sell a camera without the buyer ever *touching it* and went to another store.

Would you buy a car without test driving it? Of course not! The same holds true for a vidcam. When you are in a store pricing one, insist that the salesman allow you to pick up the camera and actually use it. There are a few solid reasons for this trial run.

First, every camera will feel different in your hands, but only one may feel *right.* You might prefer a camera that weighs a little more. The weight gives you confidence that you are holding a durable piece of equipment. However, your neighbor might like a lighter camera, which allows for quicker shooting with weaker wrists.

The weight of a camera varies from manufacturer to manufacturer. The way that weight is distributed is also varied, because some cameras are balanced better than others. There are vidcams that are made so the camera back rests on your shoulder, using a piece of plastic that attaches to the camera as a shoulder support. This will take some of the weight off your wrist. Cameras without this feature require you to press your elbows into the side of your chest for additional support. Neither way is necessarily preferable; it depends on how it *feels* to you.

The grip will also feel different from camera to camera, and placement of the viewfinder also may be a factor in the way a camera is balanced.

VIEWFINDERS

The first specific consideration when choosing a vidcam is deciding on the type of viewfinder you want to use. There are three types to choose from. Those with experience with 35-mm SLRs might be comfortable with a vidcam that has a *through-the-lens* viewfinder. It functions like its 35-mm SLR counterpart: You see the image being photographed directly through the lens, and that image is what goes on the tape.

Next is the *optical viewfinder.* Out of the three we are discussing, it is the least expensive and has the simplest mechanism. It simply shows you the direction in which the vidcam is pointing. It is not unlike a rangefinder camera, because it, too, has a slight parallax distortion: You will not get exactly what you see through the viewfinder on the videotape. For critical focusing situations, this could be a problem.

Going from the crudest and the least expensive to the most sophisticated and most costly, one comes to the elite of vidcam viewfinders, the *electronic viewfinder.* This is a miniature television built right into or onto the vidcam. It is also the most versatile of viewfinders: It shows you *exactly* what is in the frame and what is going onto the tape; it lets you see how light or dark the image is, so you can make any sort of technical corrections; and you can tell what parts of the picture, foreground to background, are in focus.

The electronic viewfinder has two drawbacks, though. Its monitor (the television set you see through the eyepiece) is in black and white, even though the camera records in color. This means that color and light corrections will be seen on the monitor as shades of gray, so it is possible at times to be off in your color corrections when in the field. At home, of course, you can color correct on your color television set so you will not run into much of a problem.

The second drawback of the electronic viewfinder is that its operation requires quite a bit of power. So when you are recording in the field, you must be extra careful about the amount of battery power you have.

As was said earlier, price and quality will play a large part in your camera selection. There are a number of vidcams with through-the-lens and optical viewfinders that give good value for the money, so do not immediately decide on a vidcam with an electronic viewfinder if it is too much of a strain on your wallet. However, if you can afford it, I recommend getting a vidcam with an electronic viewfinder for one big reason: You can play back anything you shoot with it *instantly.* This means you never have to go home from the field without knowing what you have on tape. This will save you time that might have to be spent later shooting over scenes that did not work out.

The other huge advantage of the electronic viewfinder is the ability to edit what you are shooting in-camera. This is something that cannot be done with a super 8-mm, 16-mm, or even a 35-mm movie camera. If, after viewing the tape you just shot on the monitor, you find that you do not like it, simply run the tape back and shoot the scene over again until you get something you like. It is really that simple.

LENSES

Regardless of the type of viewfinder you choose, the image will still have to be transmitted to the videocassette recorder through the video camera's lens. If the lens is made from high-quality, optically engineered ground glass and not some cheap plastic, you will get a sharp, clear picture.

The major video and camera companies, as a general rule, make good, distortion-free lenses. They have a name to uphold, so they cannot get away with making an inferior product that will produce bad publicity. The smaller companies, however, do not always have the money to perfect their lenses to the level of performance most of us would want. Given the price of a video outfit, you have a right to expect the highest-quality lens.

The vidcam itself usually comes with a zoom lens that can extend from a wide angle to a telephoto capacity. There are some vidcams that come with only fixed-focus lenses, but I would advise you to stay away from them. The various focal lengths available with a zoom will give your taping adventures a visual change of pace that you could not have with a fixed-focus lens.

There are other basic advantages to having a zoom lens. If you are shooting in cramped quarters, a wide angle lens is needed to take in all the details of the space. Then, if you decide to go outdoors to shoot an event that is occurring at some distance — a horse race, for instance — the telephoto focal length on your zoom lens can provide you with a close-up look.

The average zoom lens will have a zoom ratio of 6:1, sometimes expressed as 6×. Translated into English, this means that the telephoto focal length is six times longer than the wide angle focal length. Simple, right? Wrong.

The *range* of the zoom lens can and does differ from manufacturer to manufacturer, even though the ratio may remain constant. For example, if you take three manufacturers, they may be making zoom lenses with 6 to 36 mm, 10 to 60 mm, and 14 to 84 mm focal lengths, respectively. While each of these lenses has a 6:1 zoom ratio, the field of view of their wide angle and telephoto focal lengths is quite different.

Figure 3–1 If you're going to shoot flowers, you're going to want to get close-up to catch all the details. This would be difficult with even the best of telephoto focal lengths, because you are critically focusing; that is, your subject-to-camera distance is quite small. So, the macro mode on your vidcam will enable you to get close enough to the flowers to shoot them, while they remain in focus on the tape.

The first lens provides a wider wide-angle view than the other two. The third lens provides a longer telephoto focal length than the first two. Which lens is preferable? I would choose the 14 to 84 mm zoom because the wide angle portion of the zoom is wide enough for most shooting situations, while the telephoto end will get you much closer to your subject than either the 6 to 36 mm or the 10 to 60 mm zoom lenses. In case you are wondering, you can tell what the zoom ratio of a lens is, as well as its focal lengths, by examining the barrel of the lens.

The zoom lens on most cameras is controlled by a motor. By touching a button on the camera and keeping it depressed, the lens will zoom in or out automatically and smoothly. Some cameras even come with a variable-speed zoom, which means the zoom motor works at a slower and a faster speed. You will have greater control over zoom shots if you zoom *manually*, however. Usually, there is a bar built into the zoom lens that you can grab and then manually zoom in or out, stopping at will and moving at any speed you desire.

Remember that the zoom shot is the most overused shot in home video. I cannot emphasize enough that a zoom lens is more useful for its variety of focal lengths than for its mere ability to zoom in on a subject. I will talk more about this when we discuss shooting techniques.

There is one other feature of some zoom lenses that is of interest: the *macro mode*. The macro mode is activated by flipping a control on the zoom lens. The mode enables you to take extreme close-ups of anything from small plants, flowers, and insects to coins and stamps.

One final factor you might want to consider is whether the lens on your vidcam is interchangeable. Some camera companies making vidcams make the lens so that it is detachable; it can be replaced by a variety of lenses of various focal lengths from the same manufacturer. This adds to the versatility and capability of your vidcam.

COLOR-TEMPERATURE SELECTOR

As everyone has noted, the light is redder at sunset or dawn. Different shades of light are said to have different color temperatures. Color temperature is measured in degrees Kelvin, and it has a direct impact on your shooting.

Indoor lighting from halogen and tungsten bulbs, for instance,

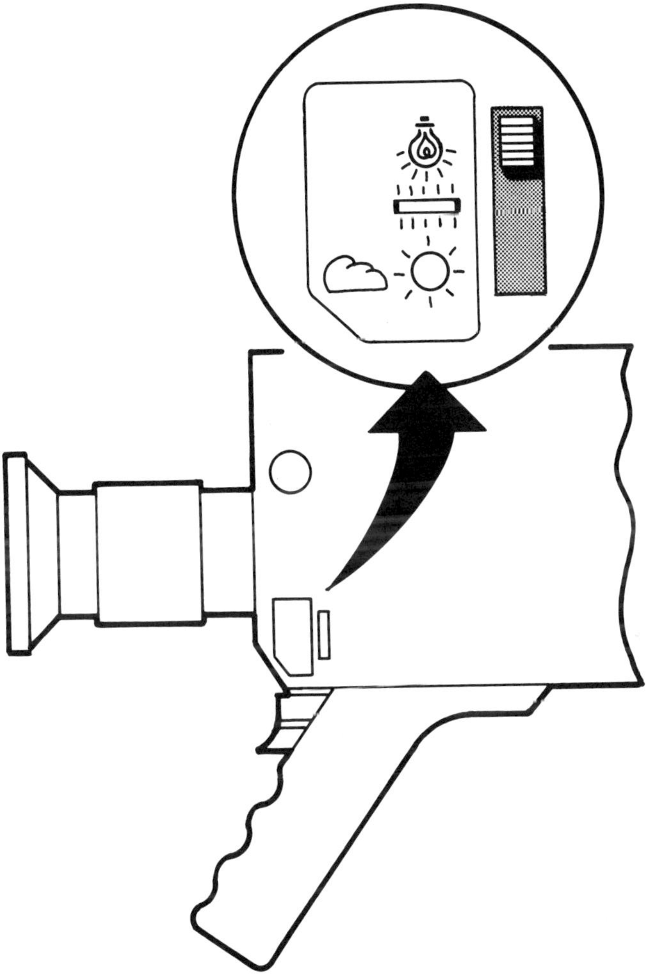

Figure 3–2 The color-temperature selector can usually be found built into the camera body right behind the lens.

measures about 3100 K, while white fluorescent lighting measures 4500 K. The light outdoors on a sunny day measures 5100 K, and light on a cloudy day registers about 6250 K. Why is all this important to you? Because there is a color-temperature filter selector built into nearly all the vidcams on the market. If you fail to use it, the lighting in your beautifully composed picture will be thrown way off.

The color-temperature selector can usually be found built into the camera body right behind the lens. Manufacturers make it easy to use; each lighting condition is normally represented by a little picture. For instance, to use the filter for daylight, set the color-temperature selector switch to the picture of a shining sun. To select for indoor lighting, throw the switch toward the picture of a glowing light bulb. Some cameras have finer selectors; white fluorescent lighting may be selected by pointing the switch at a picture of a glowing rectangle, meant to resemble a fluorescent light.

WHITE BALANCE

"White balance, when I'm using a color camera?" you may be asking. Alas, color cameras "see" in black and white, even though they record in color. The white-balance control on your camera adjusts the amount of white in your picture. If you get it accurate, you will get faithful color rendition. Why? Because every color of light is derived from white light, and every color has white in it.

Some cameras have an automatic white-balance button. All you do is point the camera at a white card or a white wall, press the button for a few seconds, and the white balance is adjusted automatically. Other cameras require you to do this manually, until it is indicated in the viewfinder that you have adjusted your white balance correctly.

Vidcams are also available that have additional color-balance buttons. This is to adjust for shades of blue and red, which are seen in the finder as shades of gray. Remember, you are focusing with a black-and-white monitor when using an electronic viewfinder. With experience, you will become adept at "reading" the light and adjusting your camera accordingly.

AUTOMATIC IRIS

The automatic iris control serves an important purpose; it allows you to set your lens opening automatically when you engage the camera, so that the proper amount of light reaches the tape. In most cases you will want to use the automatic iris, but there are two important aperture functions that require manual adjustment.

The automatic iris, which can be disengaged by turning the auto-

Figure 3–3 If your subject has a dark complexion, expose manually and do not rely on the vidcam's automatic exposure capability. Why? Because a black person's complexion requires more light to get a proper exposure. In the picture shown here, the photographer, Willie Lee Taylor, did not expose for enough light on the subject's face. Had he done so, the subject's features would have been more sharply defined. Remember, it is acceptable for the background to go a little lighter, because the subject is more important. (*Photo courtesy of Kodak.*)

matic-iris button to the manual mode, will not expose correctly for backlit situations. This is so because the meter in your camera will "read" the light for the whole scene and adjust exposure accordingly when the automatic iris is engaged. If a subject poses against a light background, for instance, the meter will average the background exposure reading with that of the subject and will set the automatic-iris control to admit less light than is needed for proper exposure. The subject, then, will be dark and underexposed, while the background will be properly exposed. By setting the iris control to manual, you can open the lens by eye until you have enough light to correctly expose for the subject. The background will become lighter, but this is less important because it is the subject that is the principal focus of the picture. We will cover backlighting situations more extensively in the next chapter.

The other function of manual mode is to allow you to fade in and out of a shot. Remember, when using the manual mode, you are manipulating the camera's lens opening yourself. To get the camera to expose automatically again, simply set it back to automatic mode.

MICROPHONES

The most overlooked factor in vidcam selection is sound quality. While any taping you do will involve communicating with pictures — and video, like its cousin, film, is primarily a visual medium — the sound that the camera's microphone is capable of reproducing can either add to or detract from the image presented on the television screen in your home.

To digress for a moment, let's consider the effect sound had on motion pictures. The producers and directors who chose to embrace it wholeheartedly were the ones who made films with sound tracks that were integral to the action on the screen. Those directors and producers who carelessly disregarded sound soon found themselves in the unemployment lines or making B films. Of course, neither predicament will befall you if you fail to use sound properly in your video productions, but your productions may turn out to be more amateurish than you would like.

Color video cameras come with a boom-type microphone that is attached someplace on the front of the camera. Some companies make the microphone so that it can extend from the camera. This serves two purposes. First, you will have better sound fidelity because you will not hear the zoom motor when it is in operation, nor will you record your own breathing. Second, by extending the microphone toward the person you are shooting, you will be able to hear the dialogue better. Your sound will be crisper and clearer.

In case you are wondering why the microphone might pick up the sound of the zoom motor and the camera operator's breathing,

Figure 3–4 It is preferable that the mike that comes with your vidcam be extendable. This serves two purposes. First, you will not pick up the noise of the camera's zoom motor running. Second, the mike will be closer to the subject, thus picking up more clearly whatever is being said. However, there are some vidcam mikes that are so sensitive that they will pick up sounds in the next room. The only way you will be able to tell for sure what a mike is capable of doing is by actually testing it out at the place where you are buying your video equipment.

you should be aware that the microphones on vidcams have a *cardioid* pickup pattern — that is, they "hear" sound in a heart-shaped pattern, so any extraneous noises around the camera will be picked up unless the mike is extended away from the camera.

You might want to consider using an external microphone when dialogue is an integral part of the scene being shot. You can purchase at any video specialty store a portable mike, called a *lavalier,* that attaches to the coat or shirt of the subject. This is the same type of mike that is used on television talk shows. Be aware, though, that when you use an external mike that connects to the microphone jack on the VCR, the camera's mike is instantly cut off. It will only begin to "hear" again when the external mike is unplugged.

Always use an earphone, plugged into the earphone jack on the VCR or the vidcam, to monitor the sound that is going on the tape. In this way you will not be surprised during playback at what you have or have not recorded.

The only real way to determine the quality of a camera's microphone, or of an external microphone, is to try it out. Record in the store, or outside it if the salesperson will let you. Perform the same test with all the vidcams you are considering. Compare sound quality during playback to decide which video camera has the best sound reproduction.

CARE OF THE VIDCAM

Caring for your vidcam, and thus providing years of trouble-free recording, is easy as long as you are not neglectful.

Lens Cleaning

Carry some lens tissue and lens cleaner with you when you go out to shoot. You can use either the lens tissue alone or the lens cleaner with it to wipe the lens free of dirt that could mar any videotape you shoot. If you use a lens-cleaning solution, be sure to apply it with the tissue or a soft chamois cloth that will not damage the lens.

It is also a good idea to purchase an inexpensive skylight filter which is then attached to the lens. This will not affect your exposure, but it will serve as a protective covering for the lens.

Lens Cap

When the camera is not in use, be sure that the lens cap is kept over the lens. Do not let it remain in your pocket, where it serves no useful purpose, even when you have stopped shooting only briefly. If, for instance, you have forgotten to put the lens cap on and, by accident, point the camera at the sun, you could burn out the camera's pickup tube.

VIDCAMS ON THE MARKET

Now that we have run through the various selection criteria you should take into account when choosing a vidcam and you know how the controls of the vidcam are operated and how to maintain it, here is a list of video cameras that are on the market. As before, this list has been broken down by manufacturer. If you have further questions about the operation of a particular unit, contact the manufacturer directly.

Sony

One of the newest video cameras from Sony, in the Beta format, is the HVC-2400. It incorporates an MF Trinicon pickup tube that Sony claims provides sharper resolution and better color fidelity.

Figure 3–5 The new HVC-2400 Trinicon color video camera offers negative-to-positive conversion capability. When used with Sony's HVT-3000 Video PhotoLab™ adapter, the camera converts color negative photography or slides into positive video images for viewing on a television or recording on a videocassette.

The camera has an f/1.4 lens with a 6:1 power zoom that has macro capability. The detachable electronic viewfinder has built into it a 1½-inch black-and-white picture tube for TV playback of taped material.

A record/review control operates with Sony Betamax VCRs to stop the recorder, play the last 2 seconds of tape on the camera's viewfinder, and then pause for the next shot. This is meant to ensure a smooth transition from scene to scene.

The viewfinder also indicates iris position, white balance, low-light situations, and tape-run conditions. You can even buy an optional cable, the VFC-6CB, that enables the viewfinder to be detached and used at distances of up to 20 feet from the vidcam itself.

Other features of the HVC-2400 include automatic fade in/fade out control, automatic or manual iris control, four-position white-balance selector, fine-adjustment white filter, built-in color-temperature selector filters, and an optional HVC-4000 remote control for access to camera controls from up to 20 feet away.

Olympus

The Olympus VX-301 tapes in the VHS format. It is light and easy to use and employs a new high-sensitivity Newvicon pickup tube that enables you to photograph at even the lowest light levels.

This vidcam comes with a 6× zoom lens (12 to 72 mm) and has a fast f/1.4 maximum aperture. Its automatic features, all with manual options, include auto exposure, auto focus, and auto color balance.

The adjustable electronic viewfinder has a large 1-inch screen. Soft-glow LEDs and indicators in the viewfinder provide a convenient readout showing record/pause, insufficient light, auto white balance adjusted, fade in/fade out, and battery power status. The camera also comes with a built-in unidirectional boom microphone.

Quasar

Quasar's latest vidcam is the VK74UE. It has a Newvicon pickup tube and records in the VHS format. The camera has automatic focus capability, an f/1.4 lens, a character generator for titling your video

Figure 3–6 Olympus's VX-301 vidcam also has a character generator for adding titles or timing and a convenient insert button which enables you to rerecord and insert a new section without erasing the previous soundtrack.

productions, a 6:1 variable-speed zoom, and a built-in stopwatch to time takes.

Canon

Another 35-mm still-camera manufacturer that has decided to enter the video market is Canon, which has come out with the VC-10A video camera, in the VHS format. It has an 11 mm to 70 mm f/1.4 zoom lens with a macro focusing ring. The zoom operates at variable speeds, and the camera itself has an automatic focusing feature so you never have to focus manually.

The electronic viewfinder positions for left or right side viewing. Inside the finder, LED displays tell the videographer if the recorder is in the pause mode, the status of the automatic white-balance control, and whether the light level is too low for getting a good exposure.

Figure 3–7 Here is Canon's video camera. It has two special features worth noting. First, there is the character generator, built in right below the manual and auto focus control. A character generator enables you to title directly onto your tapes without invoking the aid of an expensive special effects generator. Remember, though, that the more features your vidcam has, the more expensive it is. The second special feature is a 1.4X telephoto converter. Costing extra, this device extends the focal length of the camera's zoom lens to 98mm for super telephoto effects. However, the converter does not diminish the lens's high speed. It remains at f/1.4. The converter is made to be used at the zoom lens's maximum telephoto focal length (70mm), with the focus switch on manual.

In addition, the viewfinder displays how much tape has been shot and provides an indication of recorder battery condition.

To help in backlit situations, an automatic exposure lock makes it possible to correctly record foreground detail, and a fader switch provides smooth transitions from shot to shot. A tungsten daylight switch lets you adjust the camera to the color temperature of the ambient light, and another control provides color rendition adjustment.

Two other features make the VC-10A attractive. First, the lens can be taken off and replaced with any of the many lenses that fit the 35-mm SLRs that Canon manufactures. There is even an adapter available as an accessory that reduces the minimum focal length of the lens that comes with the camera to a superwide 6.9 mm for even greater coverage in tight spaces. And second, all the VCR controls, including record, playback, reverse, and frame-by-frame advance, can be controlled from the camera.

An extendable boom-type microphone, hand grip, and shoulder pad also come with the vidcam.

General Electric

General Electric is heavily into the home video field. Their top-of-the-line vidcam is the 1CVC3035E. It features a Newvicon pickup tube that works with the GE 1-inch electronic viewfinder. An f/1.4

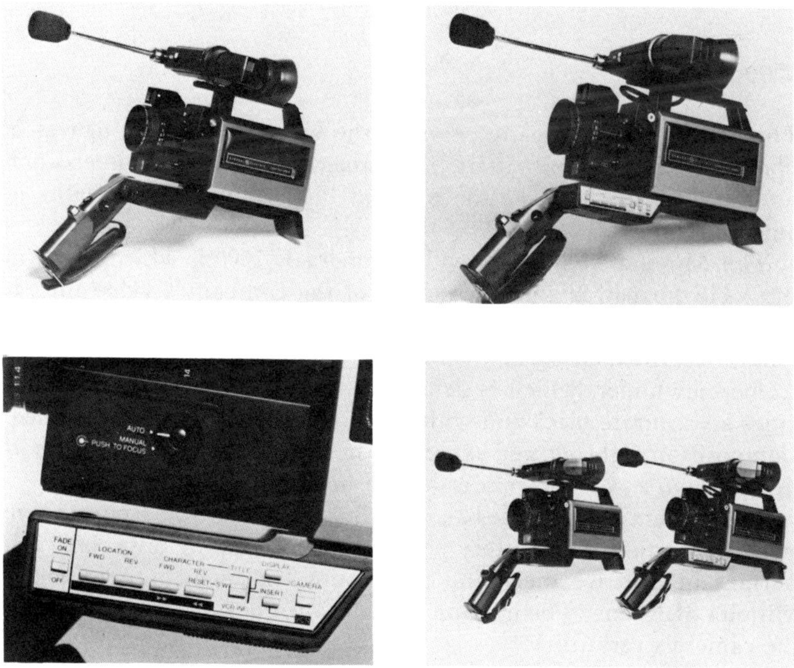

Figure 3–8 General Electric features two fully automatic focusing cameras, model 1CVC3030E (top left) and model 1CVC3035E (top right). Shown bottom left is the control panel for model 1CVC3035E which has built-in character generation and on-screen stopwatch timer as well as automatic fade-in/fade-out.

lens with a 6:1 variable-speed zoom also allows macro focusing on objects as close as 2 inches from the camera.

The camera has an automatic white-balance control, automatic iris control, an extendable-boom microphone, and an adjustable shoulder rest for added comfort. There is also a fade in/fade out switch, character generator, and built-in stopwatch. Taping is done in the VHS format.

One other gadget this camera has, like a lot of the other new vidcams, is an optional film holder. This allows the camera to be used to generate TV pictures of 35-mm color negative film by using the positive-to-negative feature of the camera. With this equipment, the videographer can view his 35-mm color negatives in full color on his own TV set *without* paying for color prints.

Minolta

This still-camera company entering the video field is one to watch. It emphasizes portability in its video cameras by employing microcomputerized chips that take the place of bigger tubes and circuits in older machines.

The Minolta MOS color video camera K-2000S, which tapes in the VHS format, is a good example of the company's video line. It features automatic exposure capability. There is a 12.5 mm to 75 mm power/manual zoom with macro capability. The vidcam's electronic viewfinder, which is detachable for special applications, provides an accurate black-and-white picture of what is being recorded. Composition, lighting, and exposure can all be monitored while shooting. Of course, the viewfinder has an instant-replay capability.

The camera has a white-balance adjustment control, a power-save switch for conserving battery power when not taping (especially important in cold weather), and a C-mount lens, which permits any Minolta SLR lens to be used on this vidcam, considerably enhancing the camera's versatility.

Zenith

Zenith has come a long way since the 1940s, when their first commercial TVs were introduced. Today, Zenith is one of the leading manu-

Figure 3–9 The new Minolta MOS color video camera K-2000S offers the latest in solid-state design. The MOS chip makes the camera smaller, lighter and more compact than tube cameras and prevents burn-in. A removable electronic viewfinder provides an accurate black and white picture of what is being recorded and acts as an instant replay monitor. Other features include a white-balance adjuster, a power save switch and a C-mount for adapting any Minolta SLR lens.

facturers of Beta-format equipment. However, while Zenith's television sets were once hand made in America, today their vidcams are machine made in Japan — and they are of uniformly high quality.

Indicative of their video line is Zenith's VC1800 Beta-format vidcam. It has a 6:1 power zoom lens, boom microphone, and color-correction filter. There are adjustments for remote/pause control, automatic fade control, and automatic/manual iris control, as well as an electronic viewfinder.

Zenith emphasizes that its zoom lens motor can be operated at different speeds. The camera's Newvicon pickup tube is especially sensitive in low-light situations. Also of special note is the automatic focus control. In the auto focus mode, a beam of infrared light is projected from the front of the camera. The beam reflects off objects at the center of the camera's field of view, and two photocells on

Figure 3–10 Zenith's new advanced VC1800 color video camera features an infrared automatic focus system. A new microprocessor-based character generator allows video photographers to add professional-looking titles to home movies. A new pick-up tube (Newvicon) provides vivid color reproduction and greater sensitivity in low-light situations with fewer adjustments.

the camera are activated, allowing the distance at which the lens is to be automatically focused to be calculated.

A drawback of the automatic focusing feature is that while it makes shooting quicker, the camera is basically taking the picture, not you. If you are going out to shoot your own tapes, you should be capable of focusing the machine for greatest flexibility in creating the effects you want.

JVC

As I have mentioned earlier, JVC is the inventor of the VHS system of home video recording. They have certainly continued to innovate with the introduction of new and advanced cameras, such as their GZ-S3U vidcam.

This camera is an incredibly light 2.75 pounds. How did they make it so light? Its picture tube is half the size of the usual picture tube in most vidcams, measuring ½ inch. There is a 1-inch electronic viewfinder, a compact f/1.2 lens, and a high-density printed circuit board using ultrasmall microprocessor chips. This vidcam is as light as most 35-mm SLRs, and the lens grip makes the camera easily operable with either hand, so the weight is even less noticeable.

The GZ-S3U has an automatic white-balance control, a 6× power zoom lens with macro setting, automatic/manual iris control (with auto lock), and an electronic viewfinder that indicates camera functions.

There is another feature that some audio buffs might be interested in. For added versatility, the built-in mike minijack accommodates either a monaural or a stereo microphone.

Any camera that weighs less than 3 pounds should be checked out before buying any other. If the quality of picture and sound resolution are as good as those of the bigger units and the price is comparable, it would be a good buy.

Sanyo

Sanyo makes the VSC450 vidcam, which tapes in the VHS format. This camera has a two-speed 6:1 power zoom and a built-in microphone on a telescopic boom.

The professional-quality f/1.4 lens lets you go from 12.5 mm at wide angle to a 75-mm telephoto with the two-speed power zoom. It even functions as a macro lens, allowing you to get as close as 7.4 mm from the subject.

The electronic viewfinder has a 1½-inch monitor that lets you see exactly what the camera lens sees. Focusing and picture composition are accomplished easily. There is a viewfinder LED feature too. It shows if the battery power is getting too low, as well as indicating whether the camera is in the pause or record modes.

Hitachi

Hitachi has come up with what they claim is a first: the first tubeless vidcam on the U.S. market. An image sensor has replaced the usual picture tube, and Hitachi claims that it is as reliable as Zworykin's invention. The camera records in the VHS format.

It is the VK-2000 color camera that has this new feature. Light, at 4.4 pounds, it has an f/1.4 lens with a 6× zoom with macro setting, automatic iris control, a unidirectional boom-type electric condenser microphone, and a 1½-inch black-and-white monitor contained in the electronic viewfinder.

The viewfinder has an interesting feature: a retractable lens for the convenience of those who wear eyeglasses. There is also a second comfort feature associated with the camera. It comes with a shoulder harness, which will help the videographer shoot steady pictures.

Akai

In its literature, Akai states matter of factly that Matsushita, a large Japanese company, manufactures the video units sold under the Panasonic, Quasar, Sylvania, Philco, General Electric, and Curtis Mathes brand names. Each company simply gives Matsushita the specifications for the features it wants in its vidcams, and the units are manufactured to spec. So while camera features do differ from company to company, it may very well be Matsushita that is actually producing the vidcam you are considering.

It was rather honest of Akai to note this practice, but, of course, the company had a reason. Akai cameras are manufactured only by Akai directly. Their vidcam is the VC-X2. It has an auto focusing feature, automatic iris control, an auto white-balance control, and auto fade in/fade out. The telephoto lens has a motor-driven 6X zoom with two speeds and a macro setting for close-ups. For in-the-field monitoring and playback, there is a standard electronic viewfinder with 1.5-inch screen.

Akai's exclusive intervalometer serves as a time-lapse device, making the vidcam useful for surveillance and security purposes, as well as for special-effects work. The time-lapse control can be set for recording periods of from 8 hours up to 11 days.

The microphone of the VC-X2 is a standard boom type, and there is a jack for a second mike. There is also a stereo mike available from Akai to produce stereo sound tracks. I should also mention that this vidcam is a light 5.3 pounds.

Figure 3–11 The Akai VC-X2 features automatic focusing, white balance, and iris control. This light unit is also capable of stereo sound.

Sylvania

Sylvania has recently gone into the home video market with its VCC116 color video camera. This vidcam features an electronic viewfinder and infrared-controlled automatic focusing, with taping in the VHS format.

Other features include a 6× variable-speed f/1.4 power zoom lens and a built-in, extendable, boom-type microphone. There is also a built-in title generator, a stopwatch, and a remote control unit in the camera that controls all functions of the VCR that are compatible with the VCC116.

Sylvania also makes a slightly less expensive vidcam, the VCC114. It also has an electronic viewfinder, as well as built-in mike, auto white balance, and backlight controls. The drawback is that it only has a 4× zoom lens.

Philco

Philco's top-of-the-line vidcam is the VCC099. Taping in the VHS format, this vidcam has a variety of popular features. There is an

Figure 3–12 One of the vidcams Philco is making is the UC112A. It has an automatic iris control that can be manually operated.

infrared auto focus control and an adjustable electronic viewfinder with a 1½-inch picture tube. The viewfinder also has contrast and brightness controls to maximize picture quality.

The f/1.4 lens has a zoom ratio of 6×, extends from 12 mm to 72 mm, and features two-speed power zoom control. There are also controls for automatic fade in/fade out, automatic iris with manual override, white-balance control, an indoor/outdoor light-filter switch, adjustable tint control, and a standby switch to conserve power.

The vidcam comes with a boom microphone, adjustable shoulder and tripod mounts, adjustable handle and grip, and a lens cap, an eyepiece cap, and a lens hood. The unit weighs 6.6 pounds.

Magnavox

Magnavox is manufacturing three color video cameras, the best model being the VR8269BK. This vidcam has an electronic viewfinder and an f/1.4 variable-speed 6× power zoom lens. The unit also has automatic and manual iris operation, auto fade in/fade out, a Newvicon picture tube to maximize brightness, and a character generator. Other features include a boom microphone, auto focusing, VCR information

Figure 3–13 Another of the Magnavox vidcams is the 8238 model. It has an automatic iris control that has a manual override, which you will find especially useful for backlit situations.

display in the viewfinder, and a built-in VCR remote switch. The vidcam weighs 5.5 pounds.

ACCESSORIES

There are some excellent camera accessories made by the vidcam manufacturers, and I think some of them are worthy of note.

Akai

Akai is marketing the VCM-1 ActiVideo Cambrace, a shoulder-mounted brace with telescoping leg that is affixed to a hip belt. Used with an autofocus camera, the VCM-1 makes videotaping easier in many otherwise difficult situations. It allows the camera operator to hold the camera steady, and it also provides the luxury of freeing both hands from having to support the camera.

Made of light-weight airplane metal, the VCM-1 weighs less than 2 pounds and can be disassembled for easy transportation. While similar, more expensive devices are used in the electronic news-gathering field, this device has been priced within the means of the home video user.

Sony

Of all the vidcam manufacturers, Sony makes more accessories than any other, among them:

Figure 3–14 The HVT-3000 Photolab converts color negatives and slides to videotape.

Remote control unit. Sony's HVR-4000 is a wired remote control unit designed for use with the company's new Trinicon HVC-2400 video camera. The unit operates recording start/stop, power zoom, and fader functions. It has a cord length of 20 feet and became available in July of 1982.

Video photolab. When used together, Sony's new Trinicon HVC-2400 color video camera and video photolab HVT-3000 provide conversion of color negatives and slides to videotape. This is particularly useful for slide presentations and for previewing color negatives.

Mounting bracket. Sony's new VCT-220 camera mounting bracket permits the attachment of Sony video cameras to the LCH-220 action pack. The action pack is a weatherproof housing for a portable Betamax. The VCT-220 allows swivel and tilting movements of the camera and provides locks to free hands for focusing. The camera's viewfinder can be attached to a control handle for easier viewing.

Special effects. Sony's new HVS-120K special effects kit is designed to provide for additional flexibility and control over the video image. The kit includes a special-effects generator and a black-and-white

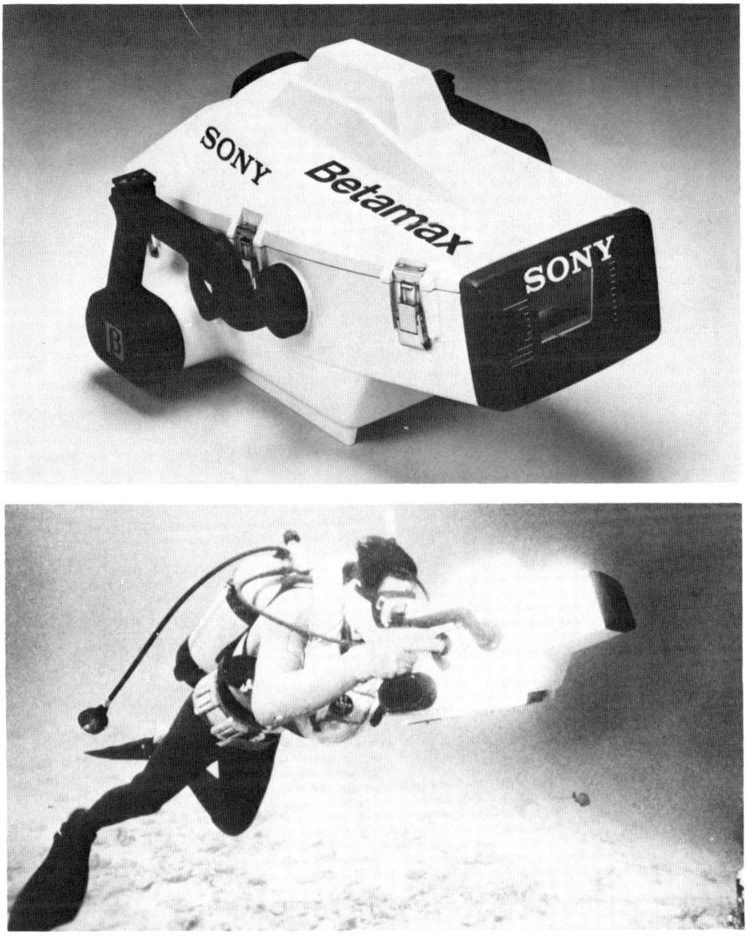

Figure 3–15 What should you do if you like to shoot underwater? The answer is to get Sony's MPK-60 Betamax blimp. This unit provides a waterproof seal for a Sony vidcam, so you can use it for underwater photography.

video camera, which, when combined with a color camera and VCR, provides such special effects as superimposing, color-keying alteration (up to six different colors), black-and-white reversal, and remote control of VCR transport functions.

Figure 3–16 A tripod, like the Sony VCT-100K pictured here, is always useful when shooting. The solid base that the tripod provides allows you to shoot steady, nonjarring images. All framing and camera movements are done much more smoothly than would be possible if you held the vidcam in your hand.

With the supplied close-up lens, camera stand, and title holder, the user can also create video titles or graphics and superimpose them over a color camera's image. The user can select any of the six colors for titles or background. The black-and-white camera image can be superimposed over a color image or shown separately if desired. In addition, titles can be faded in or out.

Microphone. Sony is also making a boom microphone that attaches to its vidcams. The Sony ECM-Z300C manually zooms from a cardioid to a shotgun pickup pattern.

Tripod. The VCT-100K is a new heavy-duty tripod made by Sony; it is for use with their Trinicon video cameras. The unit is also equipped with a shoulder-brace attachment for the camera.

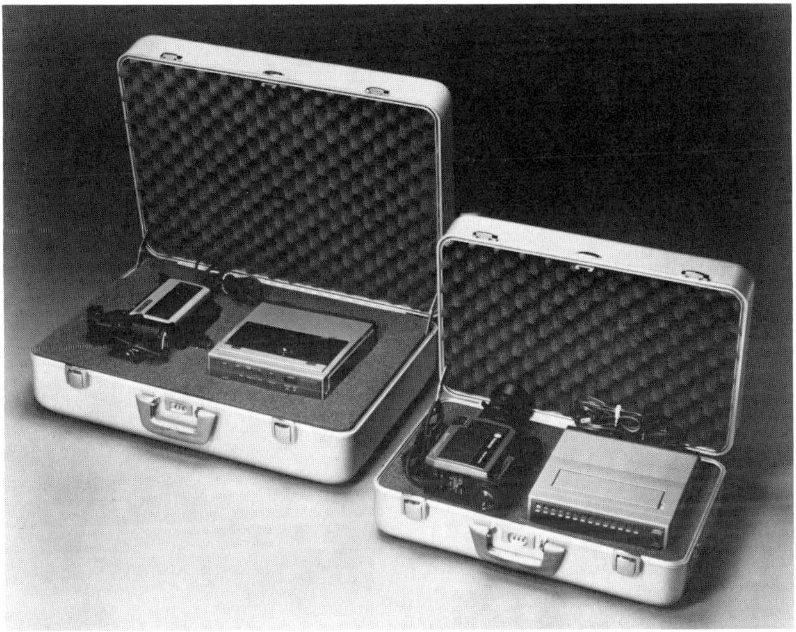

Figure 3–17 Another accessory worth noting is BMC's Halliburton custom video cases. They are made to carry portable video equipment. Construction is of prestressed heat-treated aluminum alloy to provide safety for equipment when traveling or just moving from location to location.

Cine 60

Cine 60 is an independent manufacturer of video accessories, which can be used with *any* vidcam. These include top mounting lights that fit onto the accessory hot shoe on top of many vidcams to auxilliary power packs that fit around the videographer's waist.

Specifically, the Cine 60 B-124, 12-volt Power Belt fits easily around the waist. A cable that's provided connects the belt to the VCR, which in turn is connected to the vidcam. And you can also power your top-mounted light, like the Cine 60 12-volt Powerlight, with the B-124 Power Belt.

4

Lighting

AMBIENT LIGHT AND ARTIFICIAL LIGHT

Your first consideration when lighting a scene is what kind of light you are dealing with. *Ambient light* is the dominant light source at the location where you are filming. If you are shooting outside during the day, the ambient light will be the light coming from the sun. If you are shooting indoors, away from a window, the ambient light will be the light coming from an artificial light source in the room where you are shooting. Artificial light sources include the lamps and overhead lights in the room and any supplemental lights you decide to use, such as quartz lights and photofloods.

LIGHTING OUTDOORS

As a way of approaching the techniques used in lighting outdoors, let's take two specific situations most people would be likely to face: a summer barbeque in your backyard and a trip to the beach on the Fourth of July.

If you are shooting in your backyard on a sunny day, your ambient light is the sun. However, how do you control the sun's light so that your subject is evenly illuminated? The first thing you should do is make sure the color-temperature selector switch on your vidcam has been set to the daylight mode. Once this is done and you have engaged the automatic iris control so exposure is handled automatically by the camera, you are ready to light your scene. Let's say you are shooting your father at the barbeque in his high chef's hat as he flips hamburgers on the grill and talks with the other guests.

Before you shoot, make sure that the sun is either behind you or at least at a 45-degree angle to your subject. *Never,* absolutely *never,* shoot directly into the sun. You will burn out part of the vidcam's mechanism and be left with a camera that does not work.

You also want to avoid posing subjects against a white backdrop, such as a whitewashed wall. This will throw the camera's exposure off and you will wind up with a backlit picture. So make sure that the subject is being posed against a background that is in contrast to the clothing he or she is wearing, and one that also is not white or very light.

Fill Light

Now suppose that after you follow these instructions you notice that your father's face is in shadow. Since the subject's face is the most important part of his appearance, how would you handle this situation? The answer is to utilize fill light.

Fill light is supplemental light that is used to fill in the shadowy portions of the subject. When you are shooting in your backyard or at any location near an electric outlet, you have two types of fill light to choose from. Your first choice is the simplest and least expensive; you can make a reflector board. This is a large piece of cardboard, about 30 inches by 40 inches and covered with tin foil or a highly reflective white paper. To use it, have an assistant hold the *fill card,* as it is sometimes referred to, off to the side of the camera, so the sunlight strikes it and is reflected onto the subject's face. Looking through the viewfinder of your vidcam, you will suddenly see that the face is now brightly illuminated. If it is too bright, tell your assistant to lower the card slightly, until you are satisfied with the amount of reflected light hitting the subject.

The second and more expensive way of filling in the shadowy portion of the subject is to use an artificial light source. I recommend

Figure 4–1 A reflector board can be positioned to either the left or the right of the subject at a 45-degree angle. The reflector board in this picture is made of a highly reflective white cardboard. You can also make a reflector board by attaching aluminum foil to a large board. (*Photograph by Jon Schaub.*)

a quartz light mounted on a stand, or even a top-mounted light; the necessary equipment can be purchased at any camera or video specialty store. If you use a quartz light, position the quartz light off to the side of the subject, making sure it illuminates effectively those details — such as the face — that you are most interested in. Before you take your shot, hold a white card in front of the subject and take another white-balance reading. In this way, you will be sure that the ambient sunlight will be balanced by the artificial light in the scene and that you will get a perfect exposure.

You may also want to use a blue filter over the quartz light. This will ensure that the color temperature of the quartz light is the same as that of the sunlight. Now you are assured of a correctly exposed shot.

If you are shooting in your backyard on a cloudy day, or if you are at any outdoor location where possible backlit situations do not exist, the same lighting techniques you use on a sunny day can be used.

In terms of the mood you are creating with your lighting, tapes shot on a sunny day are vibrant and convey a feeling of movement and action. Colors will be brighter. On a cloudy day you will find that the lighting is softer. It has a much more even and flattering effect and is easier to control than harsh sunlight.

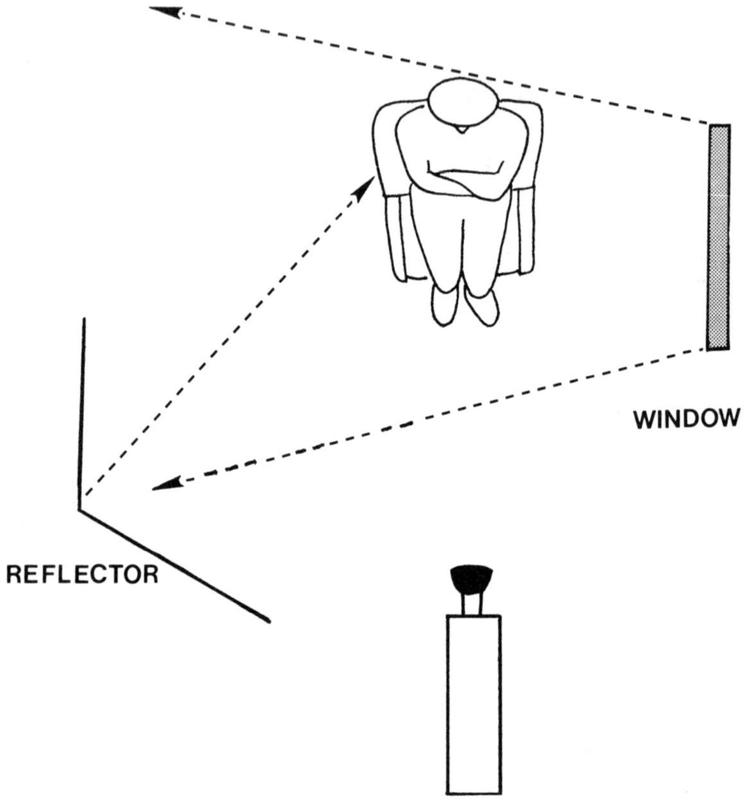

Figure 4–2 To light using window light and an inexpensive reflector, place the subject in close proximity to the window. The window light is your key light. Place the reflector off to the side of the camera's position and reflect some of the outside light onto the shadowy portions of the subject's face. The reflected light in this case becomes your fill light. This is a good way to light a subject when no artificial light source is available.

Beach Lighting

Now, it is the Fourth of July and you and your family have decided to go to the beach. You have also decided to take your vidcam and VCR along to record the event.

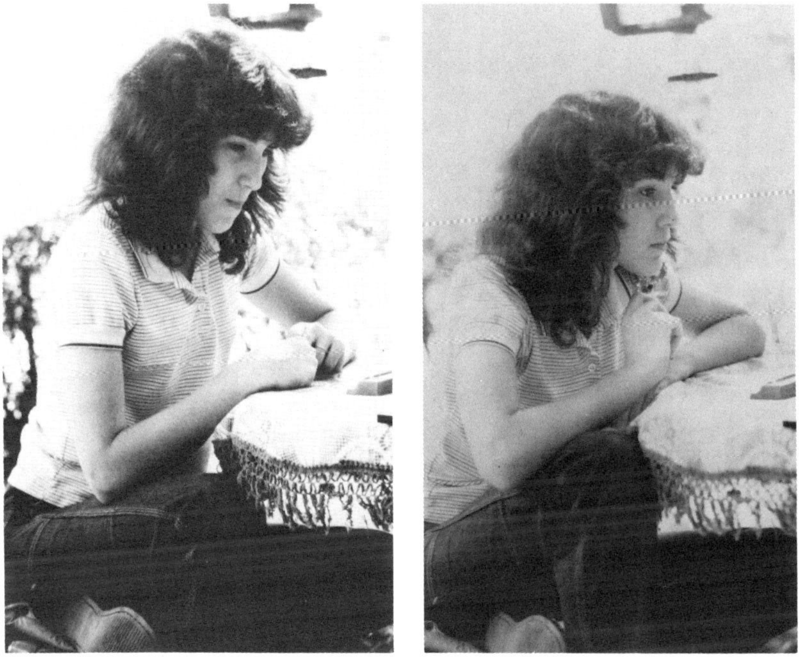

Figure 4–3 In the shot on the left, the subject is backlit and exposure was done automatically. The subject's face, the most important element in the picture, may come out underexposed in this situation. In the picture on the right, exposure was taken manually and then the shot was taken. While the background went light, the subject's face was exposed normally. Take care to do your exposures manually whenever you are taping in a backlit situation. (*Photograph by Jon Schaub.*)

If the day is sunny, you will want to avoid a backlit situation. I covered backlighting briefly in Chapter 3, but let's look here in more detail at a specific backlit situation and how you can cope with it.

Let's say that you are shooting with the sun over your shoulder or off to the side. You have already taken your white balance and your automatic iris control has been engaged. Your subject is your wife in her new and expensive bathing suit. She is moving along the water's edge collecting seashells. You look through the viewfinder, expecting to see your wife perfectly exposed. Instead, she appears as a dark image, with the surrounding sand and the water reflecting

Figure 4–4 If you decide to use your vidcam and VCR on a boat, be aware that you may encounter a backlit situation because of the light reflected off the water and the bright sky. Make your backlighting adjustments accordingly.

the bright sunlight. Even though the sun is not behind your wife, you are dealing with a backlit situation. Here is how to overcome it.

First, disengage the automatic iris control. Then open the lens all the way manually while looking through the viewfinder until you see the brightest image possible of your wife. At this point, the lens is admitting enough light to perfectly expose for your wife. Your background will have grown even lighter, but your wife, the subject of the picture, will be correctly exposed. Before you take the shot, take a white-balance reading. Now you are ready to record.

Another way to cope with a backlit situation is to handle it the same way that a still photographer would. Go up to your wife and point the lens at her face. Then disengage the automatic iris and open the lens manually until your wife is perfectly exposed. Go back to your shooting position, take your white balance, and then shoot. Once again, the background will grow lighter, but your wife's image will be correctly exposed on the tape, as well as in your viewfinder.

You might find, even if you compensate for backlighting with either of these methods, that there still is not enough light on your subject. This is where you fill in those shadowy portions of the subject by

Figure 4–5 Another backlit situation you might face is on the ski slopes. Just as with shooting at the beach, you have to expose manually for the subject and not for the background. When shooting skiers, in addition to exposing manually, you might try to shoot the skiing action from snow level with sidelighting. When the light comes in from the side, the skiers and each heap of snow will be outlined, creating a dramatic tape. (*Photo courtesy of Kodak.*)

using the reflector you have remembered to bring along from your backyard barbeque.

Position the reflector, as before, off to the side of the subject at about a 45 degree angle, and reflect the sunlight until the lighting is even on the subject's key details. Then you can take your shot.

Remember to set your automatic iris control back on the automatic mode after you are finished shooting manually. Otherwise, your exposures will be thrown off the next time you decide to use your vidcam.

Even if it is a cloudy day at the beach, you may still face a backlit situation. If the water is much brighter than the subject in the foreground, for instance, the camera will expose for the background and the subject will be a dark blur. Follow the steps outlined above to compensate. However, you may not have to use a reflector if some of the ambient light has been reflected by the sand up into the subject's face. The only way to tell is to take a careful look through the viewfinder and to judge the exposure carefully.

Paying Attention

For any lighting situation, you must pay careful attention to the way the light is falling on the subject. Is it coming from the left or the right, from in front of or in back of the subject? Look at films and videotapes to see where the light is coming from. See how the light affects a person's features. Does it flatten the nose and accentuate the jawline? Does it make the eyes look soulful and the mouth look sensual? When manipulating your lighting, take these factors into consideration. By playing with light, you can emphasize a person's good features and deemphasize unattractive ones.

INDOOR LIGHTING

To shoot indoors, you begin just as you did outdoors — by looking around to see what your ambient light source is. Let's suppose you are having a party in the living room of your house; there is an overhead tungsten light and three lamps on the end tables placed around the room, all of which are tungsten light sources, too. So your first step is to set the color-selector switch on your vidcam to tungsten.

After doing this, however, you look through the viewfinder to find that there is not enough light in the room to give you an adequately exposed picture. The answer is to have some auxiliary light sources available.

The ones we have previously discussed — quartz lights on stands, top mounted lights, and photofloods that can clip onto chairs or bannisters or be mounted on light stands — will do nicely to light your party. There are a number of lighting setups you can try.

Shooting One Person

If you are going to shoot one person, such as the musically inclined guest who's at the piano playing "As Time Goes By," you might want to use a detailed lighting setup that will require three lamps. The first is called the *key light.* This is the brightest source of illumination. It is placed at a 45° angle to your camera position and higher

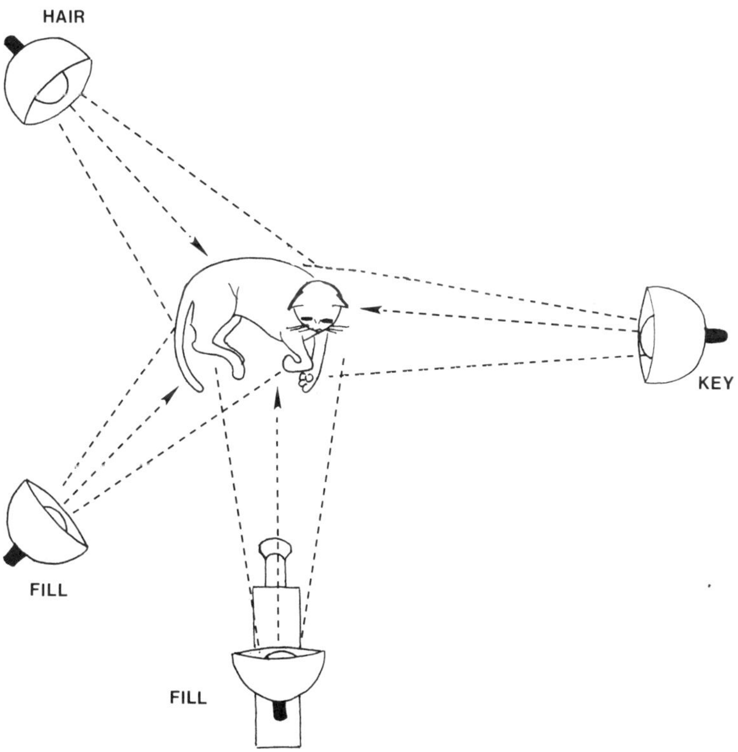

Figure 4–6 To light for one subject with artificial light, set up your key light at a 45-degree angle to the subject and higher than it. The fill light is set near the camera, and you can mount another fill light on a camera with an auxilliary hotshoe. This fill light is especially good at filling in shadows under deep-set eyes. The last light is the hair light, which goes behind the subject and off to the side. It separates the subject from the background. Remember, your subject can also be a pet, as in this diagram.

than the subject. If you are using a light mounted on a light stand, this causes no problem. Just raise the light stand above the subject's height, tighten the screws to make sure the light stays in position, and you are all set. If you are using a clip-on light, you may have more difficulty; try clipping it onto a tall piece of furniture near the camera's position or even on the aforementioned bannister if it is handy.

The second light you will be using is the *fill light.* Its function is the same as the fill light that you used at the outdoor barbeque earlier. Position it next to your camera making sure the shadowed portions of the subject's face have been filled in. If you do not have a fill light, you might want to use the reflector board.

The third light you will set up is called the *backlight,* which is sometimes referred to more accurately as the *hair light.* It is placed someplace behind the subject and off to one side. The hair light separates the subject from the background; it highlights the back, shoulders, and hair.

As mentioned earlier, it's also possible to mount a fill light that serves as an extra light source on top of some vidcams. This attaches to cameras that have a top mounting hot shoe, and serves to shine directly onto the subject's face. It's particularly good for filling in the shadows of a person with deep-set eyes.

Before you take your shot, look carefully at the subject through the viewfinder. Manipulate all your lights until you are satisfied with the way the subject looks in the viewfinder. Once you are satisfied with the way everything looks, get someone to hold a white card in front of your subject, take your white-balance reading, and you are ready to shoot. Be sure you have placed the color-temperature switch on tungsten light, because that will be the dominant light source in the room.

Shooting More Than One Subject

It is much more difficult to shoot more than one subject. People are constantly moving around, and it is impossible to use the classic lighting setup just described when many people are in constant motion. The way to handle this type of situation is sometimes referred to as *zone lighting.* In this case, you are trying to light for one portion, or zone, of the room.

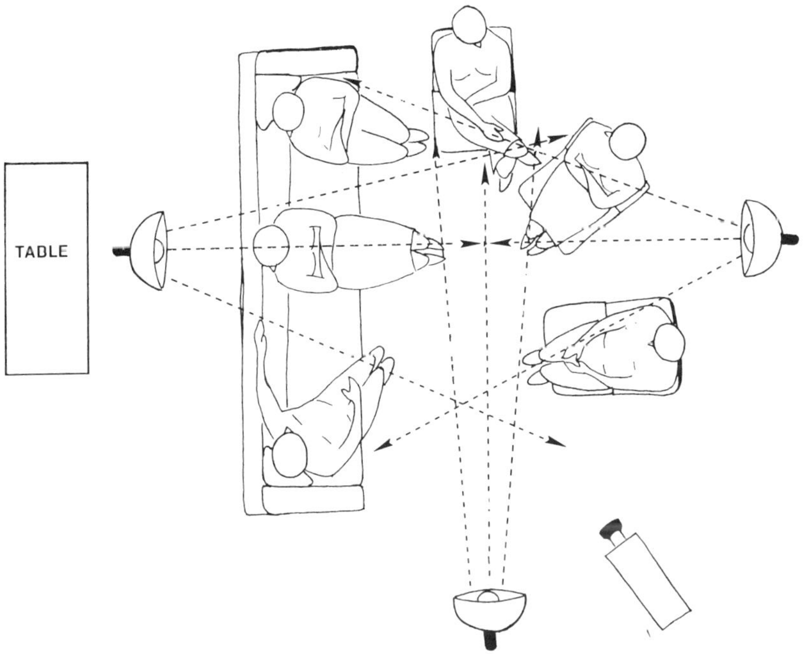

Figure 4–7 Zone lighting involves lighting the zone of a room where a number of subjects are situated. In this diagram, you can see that each light covers a specific subject. If you put barn doors on the lights, you'll be able to direct the angle of the beam. So, the more area you want to cover around the subject, e.g., a group shot, the wider you make the opening of your barn doors. The area off to the left in this diagram is unoccupied and remains dark.

For instance, let's say that everyone has decided to play a game of charades. The partygoers are gathered near the sofa, in the south end of the room. There is no point in trying to light the whole room, which includes the dining area, when the people are gathered in one small part of it, so get a couple of quartz lights or photofloods, put them on stands, and place them at the now familiar 45 degree angle from the camera, with one light next to the vidcam. Make sure there is an even amount of illumination in the picture as you look through the viewfinder of your vidcam. Once you are certain that the scene has been evenly lit, do your white balancing and you are all set to shoot.

If you find yourself in a situation where zone lighting is not possible — that is, where people are constantly moving around the entire room — you are going to have problems. Any white or very light walls in the room will throw your exposure off and create a backlit situation as you pan the camera, and moving subjects can go in and out of pools of light, making exposure inaccurate.

The best way to handle this type of situation is to greatly increase the amount of ambient light in the room. This means setting up a number of lights in the corners of the room, high on stands, and directed downward. Get some *barn doors* (available in any camera store), and attach them to your lights. The barn doors enable you to direct the light so that it does not spray out in a random pattern. In this way, each light can illuminate one part of the room, while not interfering with any other light. If more than one light is directed toward the same part of the room with the same intensity, the scene will be too bright. You can even try bouncing your lights off the ceiling for a more even lighting effect.

MIXING DAYLIGHT WITH ARTIFICIAL LIGHT

Imagine now a third lighting situation, one in which natural and artificial light are both available. Suppose, for instance, that you are visiting your parents. Their living room, where your mother wants you to do some taping, has large bay windows on the far wall. They admit plenty of sunlight, and since it is a sunny day, the room is bright and cheery.

The other lights in the room include a lamp in the far corner, where your father has sat down to read. The kitchen is off to the left, and there is some fluorescent light coming into the living room from it.

In such a shooting situation, the ambient light is the light coming in the bay windows from outside. So your first step in lighting your mother, who has just changed into her best dress, is to set the color-temperature selector on your camera to the daylight mode.

Next, adjust your white balance and make sure you are not shooting her against a light-colored background. Once you do this, you should be ready to shoot — unless you find that there are too many shadows

Figure 4–8 If you are shooting the subject *indoors* and near a window, you may find that there is not enough light coming in to get a proper exposure. The answer is to either use a reflector, or a quartz light mounted on a stand as fill light. Since there was not enough light coming into the room in the shot on the top to make use of a reflector feasible, a quartz light was used to take the picture on the bottom. (*Photograph by Gene Collins. Model is Victoria D'Angelo.*)

on your mother's face, making her features indistinct. By now you know how to solve this problem: Set your reflector up at a 45 degree angle to the camera, get your father away from his book to hold the reflector, and start shooting.

As with the fill-light situation outside at the barbeque, you can also use a quartz light mounted on a stand as your fill light instead of a reflector. However, since the quartz light is rated at a different color temperature than the daylight, you will have to compensate, or your mother's features will display an unnatural flesh tone. Use a blue filter to cover the quartz light if you use it. In this way, the color temperature of the lamp will be balanced with the color temperature of the daylight. Now, you are ready to shoot.

Depth of Field

When you start manipulating the camera's lens opening, either manually or automatically, you must be aware that you are affecting the depth of field of the picture you are shooting. *Depth of field* is informally defined as the area in front of and behind the subject you are focusing on that remains in acceptably sharp focus. This area varies,

Figure 4–9 The size of the opening of your lens will affect its depth of field. As the lens aperture is opened and the f/number goes down, the depth of field decreases.

depending on several different factors. For instance, the size of the opening of your lens (called the *aperture*) will affect depth of field. As the lens aperture is opened, the depth of field decreases. Conversely, the more the lens is closed down, the more extensive the depth of field. (Aperture is measured in *f* stops, which vary inversely to the size of the number; that is, the larger the *f* stop, the smaller the aperture, and vice versa.) Let's apply this information now to some of the shooting situations we have just been discussing.

Outdoors

On a sunny day, whether you are at home or at the beach, the lens will automatically close down, because the videotape does not need as much light to give you a proper exposure. So if you are shooting your father at the barbeque your f stop might be f/16. Behind him is the dog, prancing around, and in front of him is the picnic table, piled high with Fourth of July hot dogs. At f/16, you will have an extensive depth of field, so that the dog in the background and the food in the foreground will appear sharp and clear.

If your father is being shot on a cloudy day, however, your lens will open up more — to, say, f/11. Assuming your camera-to-subject distance remains the same, the dog behind your father and the food in the foreground will not appear as sharp, because as the f number decreases, your depth of field becomes shallower.

You will also see a big difference in depth of field shooting the backlit scene at the beach. When you first look through the viewfinder at your wife collecting seashells in her new bathing suit, she will be backlit because the lens, as previously described, closes down to compensate for the unusually bright lighting conditions. Your *f* stop might be f/16 or f/22, depending on the diameter of the lens on your particular vidcam. At this aperture, you will have an excellent depth of field. You will be able to see clearly the boat sailing in the water behind your wife and your child building a sand castle in the foreground.

Once you compensate for the backlighting, however, your lens opens up. The lower the *f* number, the shallower your depth of field. The boat in the background and your child in the foreground will not appear as crisp as they were when your wife was being backlit.

You should always be aware of depth of field because you may want to include a significant portion of the background or foreground when you are lighting your picture. In some situations, you might find that the foreground or background is more important than what the subject is doing. In other cases, you may actually prefer less depth of field to blur distracting background elements and focus attention on your subject.

SUPPLEMENTAL LIGHTING

Two types of supplemental lighting have been mentioned in this chapter: photofloods and quartz lights. Let's look briefly at the qualities of each of these lights.

Photofloods

Photofloods are merely oversized light bulbs. They are similar in filament and housing construction to the incandescent lights you find in the lighting fixtures in your home. In fact, quite a few of the photofloods available can screw into ordinary sockets, so you will not necessarily have to use a light stand. A light stand is recommended, though, for the control it will give you over all lighting situations.

Some photofloods come with their own built-in reflectors; others use reflectors that can be bought separately. The reflectors help to spread the light more evenly.

Photofloods are available in various wattages, extending from 100 watts at the low end to 1000 watts at the high end. The color temperatures of the various types of photofloods range from 2900 K to 3400 K. They can be used for either direct or bounce lighting.

There are several advantages to lighting with photofloods. They are relatively cheap to purchase, and they can be bought at many hardware, camera, video, and department stores, so access to them is not a problem. They have a fairly long lifespan, which you should take into consideration if you are going to be doing a lot of taping.

There are, however, drawbacks too. As photofloods are used, their color temperature changes. While the changes may be gradual at first, you could find yourself in a critical lighting situation only to notice that the color temperature of your photofloods has decreased. Your carefully composed and lighted picture will no longer have faithful

Figure 4–10 The Omega-reflecta 2000FC fan-cooled video light incorporates many safety features and capabilities. It also offers two 1000-watt bulbs, providing the videographer with a built-in back-up bulb or the ability to record under 2000-watt illumination. The barn doors fit over the light.

color rendition. Also, photofloods have, on occasion, blown up during use, so if you do use photofloods for your lighting, try to keep your eyes away from them.

Quartz Lights

Another lighting source you might want to consider when shooting your videotapes is quartz lights. They are similar to photofloods and household lights in that they too are incandescent lamps. However, their construction is different from that of ordinary lamps.

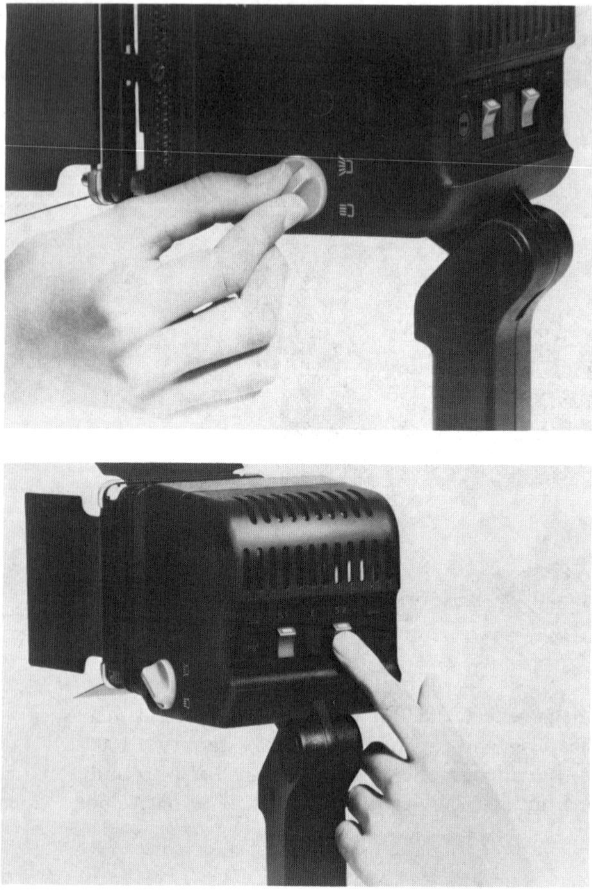

Figure 4–11 The top-of-the-line Omega-reflecta 1006FC fan-cooled video light features zoom lighting control (top) which matches the angle of illumination to the wide-angle or telephoto setting of your vidcam's zoom lens. The full and half power control (bottom), provides variable illumination intensity for more versatility.

While photofloods are physically composed of a glass envelope surrounding a filament, quartz lights have a filament that is encased in quartz or a similar substance and that is then surrounded by a vaporized halogen gas, such as chlorine or iodine.

Quartz lights come in a number of different types, with color temperatures ranging from 2900 K to 3400 K. For the videographer they are an excellent lighting source because they maintain their color temperature over their entire life span and they are extremely bright and compact. A drawback, though, is that they burn quite hot. If you touch one while it is in operation or cooling off after a shoot, you may get a third-degree burn. In addition, like their photoflood cousins, quartz lights also have blown up on occasion. Be very careful when handling them.

Quartz lights are also not as easily available as photofloods, but you can obtain a good set of them by contacting Smith Victor, one of the chief manufacturers of quartz lights in this country.

If you are interested in the latest in video lights, Berkey Marketing Companies has just come out with four new fan-cooled quartz lights. Their power ranges from 500 to 2000 watts, and all are available with the adjustable 2000-fc unit or the 1002-fc model, the latter of which is capable of 1000 watts of power. Since the former unit is more versatile, it is also more expensive. Now that we have discussed lighting, you are ready to go out and do some shooting.

5

Shooting

What is it that distinguishes the amateur from the professional videographer? The answer is the time the professional spends in framing his shot and the way he combines those shots into a coherent form using visual language.

While you may not have any desire whatsoever to be a professional videographer, that does not mean you should spend your hard-earned money on video equipment and then go out and shoot haphazardly. Nothing will guarantee the viewer's boredom, and your frustration, more than a tape of poorly conceived and ill-constructed shots.

A tape should appear to be seamless. No one shot should call attention to itself — for instance, because the camera is moving in a distracting way. You want the viewer to concentrate on the images in front of him on the television screen, not on your mistakes. Even professionals make mistakes when they go out to shoot, but they do take the time to prepare and perfect their shooting technique so that they can usually avoid hours of frustration and embarrassment later on.

COMPOSITION

All that the viewer sees when you play back your videotape is what was in the frame of your viewfinder when you took the picture. So it is important to have the subject appear as an integral part of the frame. If you study the work of some of the great film directors — such as Michael Curtiz, who directed *Casablanca*, Frank Capra, who directed *It's a Wonderful Life*, and Henry King, who directed *The Gunfighter* — you will see how the faces and bodies of the actors fit into the total composition of the frame.

Four basic compositional rules used by experienced videographers and filmmakers are (1) the center of light, (2) leading lines, (3) weight, and (4) the rule of thirds.

Center of Light

Center of light refers to the fact that the viewer's eyes are always drawn to the brightest part of the image first. So if you have a bright glare at the corner of the image you see in the viewfinder, during playback the eyes of the viewer will be led *away* from the subject to this distracting element. The viewer will have to struggle to get his attention back on the subject. So try to use light as a way of bringing the viewer's eyes *toward* your subject and not away from it.

Leading Lines

Leading lines take the eyes on a path through the frame. For instance, if the lines of the foreground lead away from the subject, the visual impact of the subject's presence will be diminished. During framing, try to look at the lines in your shot as a "map" that leads the viewer to the "treasure" — the subject.

Weight

Weight is a tool of balance in a shot. If you have a lot of visual information in one area of the frame and very little in another, the picture will be thrown off balance. For instance, having too much sky in a scene or having everything crowded into one corner of the frame will lessen the effect of the shot. The viewer will actually feel uncomfortable. How then do you make sure that a shot is balanced?

Figure 5–1 Notice how the lines in this photograph lead the eye in one direction. The eye never has a chance to wander. You should apply the same principle of *leading lines* when you compose your shots for your videotapes. Keep in mind that the more you keep the viewer's attention through the use of these visual techniques, the more involved he will be in the tape that you spent so much time in producing. (*Photograph by Mara Bovsun.*)

The Rule of Thirds

One way of checking yourself on the balance of a shot or a series of shots is by utilizing *the rule of thirds.* Freeze a picture on your television screen. Then, using a grease pencil on the screen (which can easily be rubbed off afterwards), divide the image into thirds. You will then get a better idea of how the image is visually laid out. You will see if you are crowding too much into one corner or part of the frame while not adequately utilizing the rest of your space.

The image can also be divided horizontally to check foreground and background weight. A triangular division will help you measure how much weight is in each section of the frame.

Most of us are subconsciously aware of these compositional rules. We just do not verbalize them. You certainly should not worry about them every time you make a shot. Your own internal sense of composi-

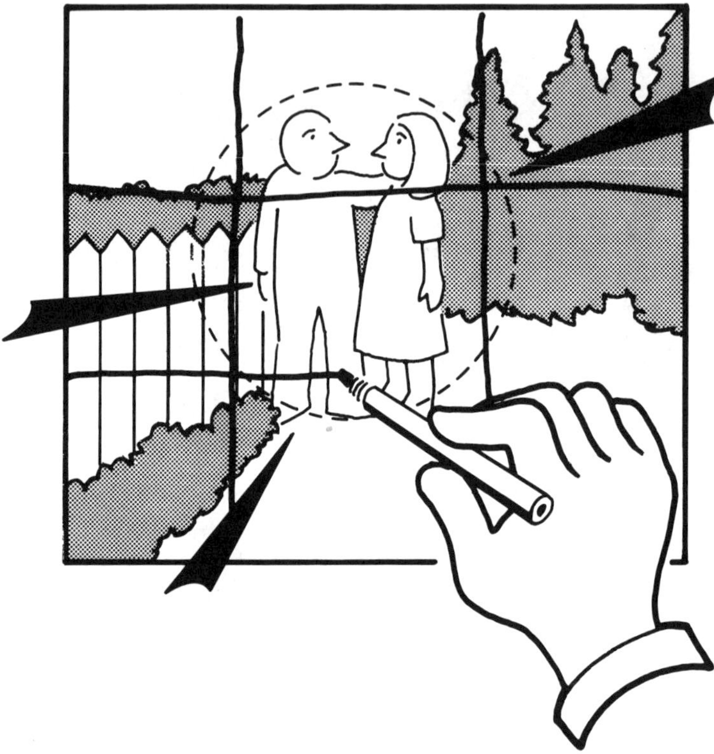

Figure 5–2 This picture illustrates several rules of composition. One way to analyze the balance and weight in your image is to divide your image into thirds (*rule of thirds*). Look also for *leading lines*, drawing the eye to the central point of interest, or the *center of light*.

tion will guide you. However, on later viewing of your tapes if you feel that your shots have an awkward appearance and you do not know what is wrong, check yourself against these simple compositional rules. You may come up with the answer to your problem.

THE SHOT

Now that you understand how your equipment works, how to light your subject, and how to frame it, you are probably chomping at the bit to go out and begin taping. However, first you need an understanding of the different types of shots and camera movements. The visual discipline you will achieve by taking the time to learn the various shots and movements and how to organize them in a coherent manner will give your tapes a new visual power.

Any discussion of shooting or shooting techniques must begin with the shot. The shot is the basic unit of any taping experience. The *shot* is defined as the tape run from camera start to camera stop; more than one shot makes up a *scene*. Within this simple definition there are three basic camera shots; each conveys a certain amount and kind of visual information.

The Long Shot

The long shot is also known as the *full*, or *establishing*, shot. It is framed in the viewfinder in such a way that you see the subject from top to bottom. If the subject is a person, in a long shot she or he will be seen from the top of the head to the tips of the toes. Other important information in the frame may include background and foreground objects, because the subject is not taking up the entire frame but rather is a small part of it.

In motion pictures and video productions, the long shot is used to set the scene for the action that follows. In a sense, it is a visual stage direction. For instance, in a dramatic production I shot called *Suspended Animation*, I opened with a long shot of a suburban home in California's San Fernando Valley. The home was the residence of a cartoonist, the protagonist in the story. The long shot simply established where the story was about to take place — the cartoonist's house.

Another example of an establishing shot is the way the first act of the popular series *Hill Street Blues* begins each week. It always opens with a long shot of the Hill Street precinct house. This tells the viewer that the scenes that follow will take place in and around the precinct house.

For shots in which you want the individual in the context of a location, use a long shot. When framing this type of shot, remember that you do not want to lose the individual within the frame, yet you do want the location to be as evident as possible.

A long shot can be the most difficult shot for your subject to cope with, because he or she has to deal with how his or her entire body looks on camera. Try to encourage your subject to interact with the environment instead of standing stiffly at attention. Use the relaxation techniques I will outline in the next chapter.

If you must, back away from the subject and use the telephoto focal length of your zoom lens when framing the long shot. This may place the individual more at ease and make him or her less aware of the camera's presence. When the individual feels more comfortable with the vidcam, you can then approach him or her and use the zoom's wide angle focal length to get a more extensive depth of field while still including the subject and a substantial amount of the location in the frame.

An effective way to use a long shot at an interesting location is to shoot slightly off the central axis of the image. In this way, you give recognition to the area and the individual. For example, if you are shooting someone next to a stream in the woods, place the stream on the center line in the frame and have the person standing slightly off to the side. Having someone at the edge of the frame looking at the scene before them will show the individual's appreciation of and involvement in the scene.

Another situation for using a long shot to best advantage might be, say, during a Thanksgiving or Christmas celebration with your family. *Don't* begin taping with shots of each family member digging into the goose or turkey. Start with a long shot of the house or apartment building where the feast is taking place. Perhaps you could also have a second long shot of the family patriarch arriving and being greeted by the hosts. However, be sure to get the initial action in a long shot — to establish the scene of the action about to take place.

The Medium Shot

The next shot you should consider is the medium shot. The medium shot is sometimes referred to as the *intermediate shot.* It is a shot where the subject is framed from the waist up. It conveys more specific information about the subject and the background, especially when compared with the long shot.

Use your wide-angle focal length to frame the medium shot, so more of the location will be shown in the frame. For instance, after the establishing shot in *Suspended Animation,* I followed with a medium shot, using a wide-angle focal length, of the cartoonist sitting at his

Figure 5–3 Compare the emotional effect of this shot with Figure 5-4 of Paul Newman. The first is a long shot from the film *The Verdict.* Notice how far back the camera is. The effect of the shot is to draw the viewer into the unreeling action. (*Photo courtesy CBS/Fox Video.*)

table drawing pictures. Then I took another medium shot of his wife, who was dusting a credenza in an adjoining room. The information conveyed was that these people were in the same house but in different rooms, an important consideration when the cartoonist has a stroke later in the production. In both medium shots, you could pick out details of both the subjects and their backgrounds as the viewer's attention was drawn into the unreeling action.

One interesting thing about the medium shot is that it produces frustration in the viewer. It hooks him and makes him want to get closer to the subject, be it Aunt Sally at the family feast or Captain Furillo on *Hill Street Blues.*

The Close-Up

The close-up is perhaps the most overused and hackneyed shot in the amateur and professional's video repertory. The close-up covers

Figure 5–4 In this second shot, director Sidney Lumet has moved his camera in for a medium shot on Newman. Now, we can see more detail on his face, the clothing he is wearing, even the ring on his hand. The medium shot draws us in closer to the on-screen action. (*Photo courtesy CBS/Fox Video.*)

the head and shoulders of the subject. The background is totally subordinated to the subject in the foreground.

More than any other shot, the close-up reveals a person. The camera picks up every nuance of expression and acts like an optical window to the soul. One close-up is worth three times its weight in medium and long shots. That is why you should use the vast power of the close-up sparingly.

On television shows, the close-up is used more than any other shot. Perhaps long ago some filmmaker discovered that a close-up of an actor or the host of a TV game show would fill the entire screen with the subject's face and decided that this was a desirable state. What that filmmaker did not know was that because of the very intimate nature of television, the close-up has even more power than it is capable of on a movie screen. And so, because it is the easiest shot to frame, and because it has such visual power, the close-up became overused on the small screen.

What the videographer should be aware of is that by cutting back on his close-ups, when he does use them they will be more significant and have the power they were intended to have in the first place, when D. W. Griffith invented the shot. Using too many of any one shot only dilutes your shooting technique. When shooting my own material, I have always tried to use my close-ups sparingly.

The Extreme Close-Up

The extreme close-up concentrates on a specific aspect of the subject's appearance; e.g., eyes, nose, mouth, hands, etc. Because the frame will be filled to bursting with a specific aspect of the subject's appearance, it carries even more visual power than the close-up and should be used *only* for a specific purpose.

Figure 5–5 In this long shot of model Victoria D'Angelo, you can see the subject in the context of the location. You can clearly see the fire escape she is standing on, the buildings of Manhattan in the background and, of course, the subject herself.

Figure 5–6 The medium shot emphasizes the subject and subordinates the background, as in this picture of Victoria. The background is not as clear, and you can see more details of the subject's physiognomy and dress. There is also more texture in a medium shot, more of a "feeling" for the subject.

For instance, you may want to use it to call attention to a particular physical characteristic of your subject. You have to decide which features to concentrate on, though. For instance, if the eyes show character, shoot from slightly overhead so that the eyes will completely dominate the picture. Stress an acquiline nose and a "determined" jawline by shooting in profile.

Whatever features you decide to emphasize, try not to shoot head on; this tends to flatten a person's features, especially when using the wide-angle focal length. Remember that most people are sensitive about the way they look. If I shot my Uncle Harry's large nose in extreme close-up and then played the tape back with Harry in the room, I would be lucky to escape with my life, unless he had a sense of humor (which he doesn't).

Use restraint and control when using the extreme close-up, as well as the close-up. Especially reserve the extreme close-up to convey to the viewer a piece of visual information that might be lacking in the other shots.

BASIC CAMERA MOVEMENTS

Once you master these four types of shots, you are then ready to begin experimenting with body and camera movements. There are two basic body and camera movements that can be accomplished with your vidcam: the pan and the tilt.

The Pan

Like most elementary body and camera movements, the pan can be done with a tripod-mounted vidcam or by using your own two hands to move the vidcam. *Panning* involves moving the vidcam from left to right or right to left. It is a shot that is used to give flow or movement to your tape; the scene might remain hopelessly static without it.

The pan, because it is so easy to do, also tends to be an overused shot. You should pan when you want to include some additional information in the frame; you should not pan indiscriminately.

For instance, in *Suspended Animation*, the cartoonist has his stroke and his wife takes a job at a local hotel to make ends meet. This scene opens on a porter pushing bags through the hotel lobby. The camera pans with him as he goes by the check-in desk, where the camera movement stops. Behind the check-in desk is the wife, working with an adding machine. What the pan did was to help get into the scene in an effortless, fluid, and natural manner.

At the family feast I mentioned earlier, you would also have an opportunity to use the pan. Let's say that on one side of the dining room table is Uncle Moe, cutting the roast beef. On the other side of the table is your brother Jules, anxiously waiting to be served his portion. (Jules has not eaten in 3 days, in preparation for this meal.) A pan from Moe to Jules will establish that one action, the carving, is giving way to another, the salivating. You do not need dialogue to convey the meaning. The action will do it for you.

The Tilt

The second basic body and camera movement is called the tilt. A *tilt* involves moving the camera in a vertical line from the top of

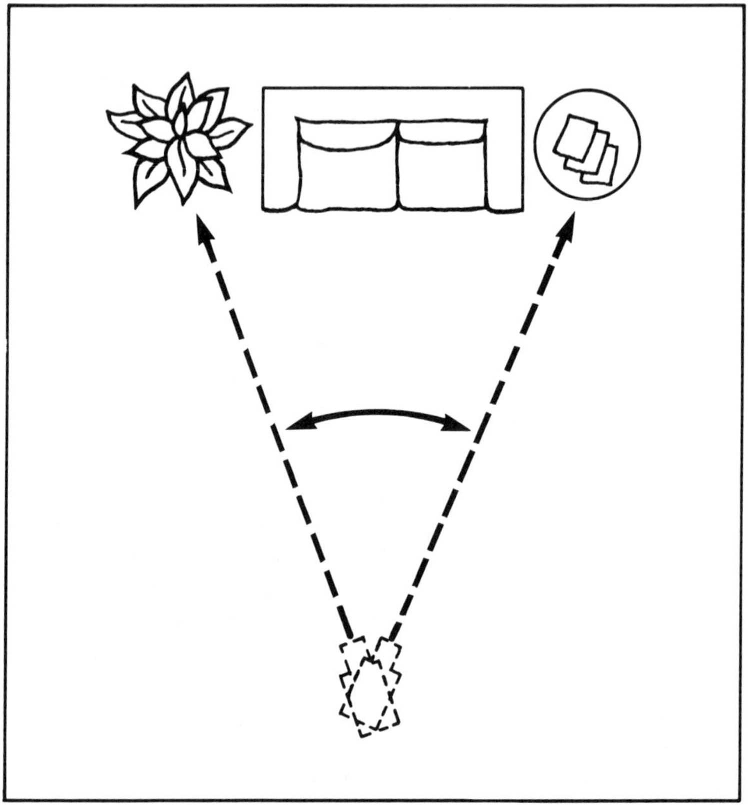

Figure 5–7 *Panning* is a shot that is used to give flow or movement to your tape.

the subject down or vice versa. For instance, after the pan got us into the hotel scene in *Suspended Animation,* I cut to a medium shot of the wife pressing the keys on an adding machine. I then cut to a close-up of her fingers playing across the keys, and slowly, I tilted up to her face as she stared at her fingers, aware of the fact that her husband the cartoonist had been denied this basic pleasure. Again, no dialogue was necessary. The tilt and the physical action said it all.

Getting back to our holiday feast, let's say you want to tilt from Jules, digging into his heaping plate, up to his still salivating mouth.

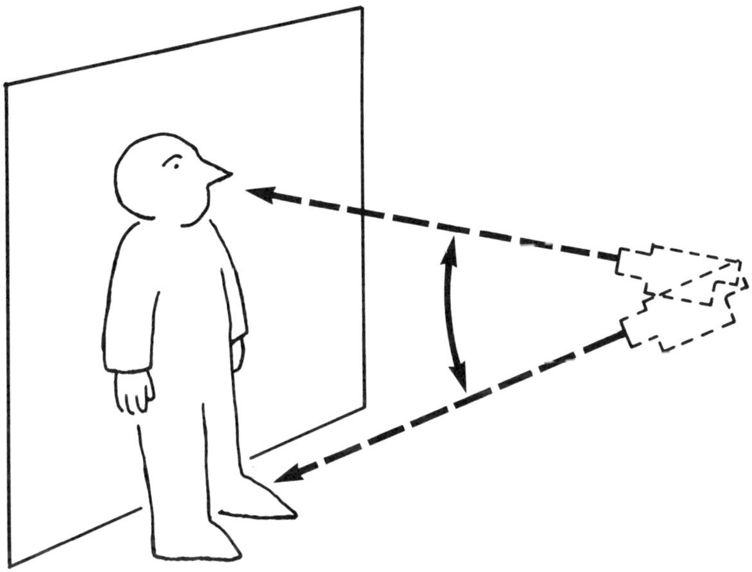

Figure 5–8 A *tilt* involves moving the camera in a vertical line.

Start your tilt on Jules' fork digging in; then follow the fork slowly up to his mouth. What you have just done is conveyed the visual information that not only is Jules' mouth waiting for the roast beef, but his hand and arm are also expressive of his expectation. Now this may be an exaggerated situation, but it gives you the idea of what you can do with the tilt.

When you use either movement, the tilt or the pan, on a tripod or with a hand-held camera, keep the camera moves slow and smooth. Tape will be wasted if you go too fast, because the image will be blurred.

OTHER CAMERA MOVEMENTS

There are three other camera movements that you can execute: the dolly, the tracking shot, and the boom shot.

Dollying

The word *dolly* has two meanings: It can describe a kind of camera movement, or it can be the name for the instrument producing the motion. First, we will discuss dolly as a camera movement.

A dolly shot is simple: The camera *physically* moves closer to or away from the subject. Do not confuse this with a zoom shot, where the lens *optically* dollies in. Remember, the camera is *physically* approaching or retreating from the subject.

Why put the dolly shot into your visual repertory? Because it adds further visual information about the subject and the background and how they interrelate.

At the feast we have been attending, suppose that the family is saying grace before the meal and you want to get it on tape. You could use a long shot and get everyone in the same shot, but that would be too static and would show little detail. Instead, take your camera, jam your elbows into your ribs for support, and slowly dolly

Figure 5–9 A *dolly* shot involves physically moving the camera to or away from the subject.

Figure 5–10 Slik Tripod has built a video dolly, seen here in the lower part of the frame, with three of Slik's video tripods above. The video tripods have model numbers 112V, 312V, and 212V, while the dolly's model number is U700.

down the table. With one continuous shot you will be showing each person saying grace in their own way. In one shot you will have both intimacy and movement.

The second type of dolly shot involves mounting the camera on a movable platform, or dolly. The platform then moves to or away from the subject, or in some cases around him. This type of dolly shot can show the "different sides" of the subject for striking effect.

In the motion picture *Judgment at Nuremberg*, for instance, director Stanley Kramer chose to have his camera dolly around Montgomery Clift, who was testifying at the Nuremberg trials. The camera made a 360-degree circuit around Clift, who could not escape the camera's probing eye; he appeared as a man caught in the vise of his personal problems, as witnessed by the camera and courtroom spectators on all sides.

Tracking

Another type of camera movement is the *tracking shot.* During a tracking shot, the camera physically moves along by the side of a *moving* subject. In effect, the camera is tracking the subject.

The tracking, or trucking, shot, as it is sometimes known, can have a claustrophobic effect, because it never lets the subject get out of the frame. It is also a difficult shot to master because you have to keep the subject in the frame as you move along beside him. In effect, you have to follow the rhythm and timing of the subject's movements, which will not always be in your control.

Going back to *Suspended Animation,* the cartoonist is confined to a wheelchair during his rehabilitation. Several times his movements are followed in a tracking shot as he wheels the wheelchair around with his one good hand. The effect is to show a man trapped within the confines of his handicap.

It is the easiest thing in the world to overuse the tracking shot at a family gathering. If the participants are moving about after dinner and the camera is tracking them without any real purpose, you wind up with a tape in which the subjects move aimlessly, the camera moves randomly, and everyone gets dizzy watching it.

Use the tracking shot when the situation requires it or the story

Figure 5–11 Another type of camera movement is the *tracking* shot, where the camera physically moves alongside a moving subject.

Figure 5–12 The football game after the holiday dinner is an ideal time to use the tracking shot. By following the flight of the ball, you can create tension in your tape. Will the receiver catch the ball or not? (*Photo courtesy of Kodak.*)

will be enhanced by it. For instance, let's say that after Thanksgiving dinner the family decides to play touch football on the lawn, and you decide to cover it with your vidcam.

When a pass is thrown, you can follow the ball's flight using a long shot, which will take in more information than any of the other shots, but it is going to be a boring shot because there is no one person you are focusing on. If you pan from the ball in the air to the receiver, you set up a certain amount of tension, because you are not sure if the ball will wind up in the receiver's hands or not. Instead, ask yourself this question: Who is really the principal character in the scene? The answer is the receiver. Will he catch the ball or not?

Providing there are no toddlers underfoot and you can move over relatively smooth ground, try a tracking shot. Start shooting the receiver *before* the ball is passed. Follow the receiver as he moves back to catch the pass. As the cameraman and as a viewer, you only know what you see in the frame. You will not know if the receiver will catch the ball or not until the very last second when the ball appears in the frame. You have therefore set up the most suspense you could in the shot.

When you play the shot back on your videocassette recorder, you can do it at an even slower speed to heighten the suspense, just like director Hugh Hudson did when he slowed the racing sequences in *Chariots of Fire* down to slow motion.

All the camera movements I have just described can be done with a hand-held camera. If you do the shots this way, use a body brace or hold the camera close to your body for support, and if possible, hold your breath during the shot, since holding your breath minimizes extraneous camera motion.

One other point. While most of us do not have access to the professional dollies I mentioned earlier, improvisation will sometimes work. Get a shopping cart or a child's wagon. Get in with your vidcam and VCR and have a relative or friend push and pull you around while you are doing your dollying and tracking shots — crude, but effective.

The Boom Shot

There is a third type of body and camera movement I have not mentioned yet. It is called a *boom shot,* and it can only be effectively achieved using a boom, which is like a crane that moves up and down in an arc with the camera attached to it. It is a strange sort of shot, almost like a mechanized tilt. It can be extremely effective at showing, for instance, a person's loneliness.

In the motion picture *High Noon,* director Fred Zinneman used an unforgettable boom shot. He started with a close-up of Gary Cooper, playing Marshal Will Kane. Kane is standing on a street, his agonized face filling the entire frame. Slowly, Zinneman boomed up, gradually showing more and more of the background. Finally, we see that Kane is all alone on a deserted street in a deserted town, ready to face four deadly killers who have sworn to shoot him.

It is difficult to do a boom shot without a boom, but you can simulate one. Every family seems to have its loner, the person who sits off to the side while the main action is going on. Suppose that after the holiday dinner, while most of the family is playing touch football, the loner is sitting off to the side, reading a book. Bend down and focus on him, his body and face filling the frame. Make sure the football game is being played directly behind him. Now slowly stand, or "boom," up. You will suddenly see the game appear in the background, while the loner remains in contrast by himself and uninvolved in the foreground.

ZOOMING

There is one last shot we should discuss, namely, the zoom shot. I have left this shot for last because, frankly, I do not think very much of it. It tends to call attention to itself, and any shot or camera movement that calls attention to itself diverts the viewer from what is most important — the subject. Too many people purchase a vidcam with a zoom are constantly zooming in and out, thinking this is creative camera work. It is not!

The best way to use your zoom lens, as I have stated before, is to take advantage of its various focal lengths. The normal focal length of the zoom lens approximates what the eye sees normally, while the wide-angle focal length takes in a much larger field of view. A telephoto focal length allows you to get close to your subject without physically doing so, simply by adjusting the lens. The zoom lens, then, allows you to frame your shots at a variety of focal lengths without changing lenses.

If you must use the zoom lens for a zoom shot, you should focus the lens in advance; in this way, as you move through the various planes of focus, the image will remain in focus. How do you do this? Set the lens at the telephoto focal length and focus; then zoom back to the wide-angle setting. As you *slowly* zoom in, the image will remain in focus throughout the range of focal lengths it passes through.

CAMERA ANGLES

One final thing you should consider when shooting is to change camera angles from time to time. Nothing is more boring than a scene,

Figure 5–13 At any family gathering that you decide to tape, you are always going to run into that one fellow who decides to mug for the vidcam and spoil that perfect shot. Just tell him not to act for the camera, unless that is what you want. The advantage of shooting video, of course, is if someone *does* ruin a shot by making a face at the camera, all you have to do is tape right over it with another shot more to your liking. (*Photograph by Nathan Alper.*)

such as the family feast, seen over and over again from the same angle, regardless of the kind of shot you frame.

Change camera angles so the scene has more visual "punch." However, if you are shooting a subject from the front and you suddenly change angles and shoot him from the rear, the viewer may become disoriented from this "jump cut." Instead, make believe that there is an imaginary line running through the subject. Shoot from any place on one side of the line, but not from the other side. If you stay to one side of the line, you will never have to worry about jump cuts, and you can then shoot the subject from any angle. Now you are finally ready to go out and shoot with your vidcam.

6

Video Portraits

Give a young child a crayon and a piece of paper and he or she will do what his or her ancestors have done for millions of years — draw you a portrait of a family member or friend. If you look closely at the child's portrait, you will see that it emphasizes the subject's most obvious physical characteristics. The man with a small mustache suddenly has a handlebar, while the woman with a slightly larger than average nose now has a desperate need for plastic surgery.

When you are shooting someone with your vidcam, you have a harder task than the child artist. You must be more subtle in your taping, yet more revealing. You should be attempting to capture the *essence* of the person when you shoot him or her. This essence is really the individual's personality, the one thing that makes the person different from every other living thing.

LOCATION

One of the ways you can help show the personality of your subject is to choose an appropriate location in which to do your taping. The location can be an expression of an individual's status in life, of his connection to home and family, or of anything else he feels may be special about his ideas and feelings.

At the Office

Perhaps you have decided to shoot a friend or relative at his job, or maybe you are called in professionally to take an occupational video portrait — a tape done of a person while he is working. You will find that an occupational video portrait can be particularly revealing of someone whose identity is intimately linked with his career.

For example, a judge can be shot in his chambers, with rows of leather-bound lawbooks in the background. A doctor can be taped treating his patients, while a railroad executive would be thrilled to be shot behind the throttle of his company's new metroliner. And

Figure 6–1 Occupational videotapes can be done of individuals in professions as diverse as doctors and truckdrivers. You can even shoot a farmer, like Skip Neumann, who is pictured here. Neumann farms acreage in Vaughn, Montana; he is posing on top of his new tractor. (*Photograph by Mara Bovsun.*)

what better way to capture a farmer than at the wheel of his tractor, plowing his fields?

Aside from sleeping, we spend more time at our jobs than in any other activity. This time has much to do with our definition of who we are and how we see ourselves.

In the Home

Tapes shot in the home have a feeling of intimacy about them. Home is one of the most important places in our lives. Many people consider their family to be *the* most important facet of their lives. They take pride in and have a great sense of security in their home environment. Whether it is a house or an apartment, home is a place of special significance.

Each person in a family has his own place within the home: the father, his den; the kids, their playroom; and the mother, the living room, which she has taken the time to decorate in colors and patterns that say something special about her innate sense of beauty and style.

Figure 6–2 Holidays, like Easter, are a great time to shoot the family. You might want to shoot the children during an Easter egg hunt, as in the picture shown here. (*Photo courtesy of Kodak.*)

By shooting family members in their own special places in the home, you help to show a part of their personalities that might have remained unseen in a plainer background.

Another place you might want to shoot at home is in the backyard. Tapes of children on swings or dad in his apron at the barbeque will make for interesting viewing. Remember, home and environs provide a wealth of taping opportunities.

During any taping session, use the location as a statement of how your subject sees himself or wants the world to see him. Sometimes the latter is more important, so shoot a sophisticated city dweller coming down the steps of an elegant hotel. Tape an inveterate traveler in the bow of a boat looking pensively off at the horizon.

Taping someone on top of a tall building with a city arrayed majestically behind him can give the feeling that they are on top of the world. And a child dressed in his little league uniform, bat in hand as he goes inside a major league ballpark will create premonitions of things to come. Locations should reveal rather than conceal the true character of the subject.

RELAXING THE SUBJECT

Many people who are confident in most circumstances get insecure once they step in front of a camera. This holds true for children as well as adults. Part of your job when you are taping is to help the individual relax in front of the camera.

Children, for instance, should be approached with love and respect, *not* condescension. Crouch down when speaking with a child, so you are physically on his level. Make conversation that flows. Find out what he likes when eating out (a frequent response will be hamburgers) and what he saw the last time he went to the movies ("I saw *Superman* four times!").

By speaking with children before setting up your vidcam, you will get them to trust you and thus to relax. Have patience, though. It may take many conversations before the child lets down his guard and acts natural. Children are incredibly perceptive people. If they believe you are allowing them to be themselves, and not forcing them to be someone you would like them to be, you will usually have a willing subject.

In getting along with children, you have one great ally — the child's vanity. Readily compliment a child who is awkward. Try to make children feel good about themselves. Play a game of hide and seek

Figure 6–3 Shooting children as they interact makes emotionally effective video. Try to relate to children on their own level, but never condescend. Get the children that you are shooting to feel comfortable in front of the camera. (*Picture courtesy of Kodak.*)

to relax the overactive child. If the child is afraid of the vidcam, let him touch it and perhaps take a peek through the viewfinder. If you are in the process of trying to shoot a child who has trouble looking animated, ask him if he can see himself in the lens. This question will produce a curious, attentive expression.

Toddlers

One of the biggest problems the videographer will encounter is shooting babies. How do you get them to stay in place long enough to tape them? They will constantly crawl out of the frame, and even if you follow them with the vidcam, your visual technique will become overactive, with constant panning. So how do you get your shots?

Try waiting until the child gets hungry. Sooner or later, he has to eat — and you will be there, ready to shoot. Check out the child's eating area ahead of time. Be sure the background offers contrast to the child's coloring and dress. Remove distracting objects from the background and foreground. Move the eating table slightly if its background is too distracting or the wrong color. Moving the child from his regular eating area rarely works. Most kids like to stick to their daily eating routine. Changes upset them, and the last thing you need is a crying child to cope with.

Adults

Adults also may feel uncomfortable in front of the camera. They may be self-conscious about their looks, clothes, even their hair style.

Figure 6–4 When you are shooting a tape of a family, you may find that the mother is a bit more traditional than some. She may feel more comfortable being shot in *her* kitchen, like the woman pictured here. The only way to know for sure where the subject will feel more comfortable being taped is to interview the person before shooting even one foot of tape. Notice, too, in this shot that the videographer forgot to put a little pancake on the face of the subject, which would have cut down on the glare of the lights. (*Photograph by Gene Collins.*)

Take time to reassure them about their appearance. Be sensitive, showing them that you are interested in who they are and what they do. If you are shooting an executive, treat him like one, because that is what he expects. The actress or dancer will love to talk about her art; the housewife, her family.

Whenever you are taping someone, discuss with them the feelings you are trying to convey through your visuals. Give your ideas, but always be a *listener* first. Most of your understanding will come from the "lost art" of *hearing* what the individual has to say. These communication techniques apply when taping family members as well as others. Just because you are related to your subject, do not take him for granted. Sit and talk with him too.

Try to make the individual part of the taping process. Discuss the technical aspects of what you are doing. You do not have to go into elaborate detail. Just give your subject an idea of how it all works. By sharing your knowledge of videotape recording, you build trust and confidence.

PROPS

Even when you feel you have managed to relax the subject, you may still see a tense person when you look through the viewfinder. One way to loosen the subject up is to use props that are indigenous to the place where you are shooting. Having the hands hold something gives the individual something to do while the tape is being shot. It will relax him.

Recently, I was asked to do an occupational video portrait of a local merchant who happened to be a furrier. He wanted the tape to reflect his place in the business world. The idea was to follow him around for a day — to see what he did at work and to shoot him with his family. One of the locations used was his new shop.

Using long shots, I shot him in his work apron, standing in front of a sewing machine he used to sew the mink skins that became fur coats. He talked about how he had come to this country at the age of 12, worked in a bakery until he was 17, and saved up enough money to open his own business.

However, looking at him through the viewfinder, he appeared ill at ease; he kept fidgeting. I suggested that he cross his arms on his chest in a pose meant to show the confidence of a man who had

Figure 6–5 One of the places that I shot the furrier, whom I refer to in the text, was in his home after a day's work. Since the subject felt the most comfortable being posed in the recliner in his living room, that is where he was taped. I also used a cigar as a prop to further relax him. Unfortunately, I forgot to tell him to remove his glasses, which reflected the lights in a distracting manner. Still, you can see how relaxed he was.

his chest in a pose meant to show the confidence of a man who had been in business for himself for over 35 years. Still, he appeared uncomfortable; his arms and legs were stiff and awkward.

Because his hands were used to manual labor, I asked him to hold a tool of his trade. He chose a pincher, a simple metal device much like a scissors that is used to stretch the mink skins flat on a board. Once again, I looked through the viewfinder as he spoke of his many years in business for himself. Holding a tool he had used for all those years, he now felt comfortable. He relaxed, spoke easily, and was very happy with the finished tape I presented him a few weeks later. Props add another visual clue to the person's true identity.

When using props, try to match them to the location and the individual's background. They can range from the mundane to the inspired — couches, chairs, PingPong paddles, pool sticks, or vacuum cleaners in the home and backyard; benches, radios, sandwiches, sombreros, and Frisbees on picnics and at the beach; fences, saddles, horses, ropes, and wagon wheels on ranches; compasses, sextants, sails, and anchors on boats; baseballs, footballs, and soccer balls on playing fields — all can be used as props. Sometimes a prop may not even be seen on the tape; still it could be the missing ingredient in creating a great tape from a situation that might have appeared lackluster without it.

In another case, I was called in to do a tape of a middle-aged woman, a buyer for a local department store. Her family had recently purchased a home deck and she wanted a tape of herself to play on it. Since she did not have a vidcam or a desire to own one, she hired me to do it.

During our preliminary conversation, I discovered she had a loving family and wanted to be shot in the living room of her home. I placed her on a love seat by the window. She held her hands in her lap. I decided the most effective shot would be a close-up.

As I looked at her through the viewfinder, she appeared to be uncomfortable. She had high cheekbones and a strong jaw line, which gave her face a hard look; this was only heightened by her nervousness. She was constantly shifting her hands.

While she spoke about her experiences as a department store buyer and her love for her family, I ran the tape, changing camera angles occasionally. I had to stop the tape a number of times because she moved in and out of the frame as she fidgeted.

I knew I had to relax her to finish the assignment, so I looked around the room for a prop to keep her hands occupied and perhaps to help relax her. There was a tennis racquet on the wall in a far corner of the room. I asked her what it was doing there, because it seemed out of place in the handsomely decorated room.

With a trace of embarrassment, she said she was a tennis champion in high school some 30 years ago. She had started out to be a physical education teacher, but had gotten sidetracked. The racquet, she said, was "sort of a momento."

This is great, I thought; if only I could get her to show that same emotion on camera. I took down the racquet, gave it to her, and said to hold it out of camera range. She thought this was ridiculous, but did as I asked.

When I taped her again, she was a different person. I got her to speak about her hopes and dreams as a youngster, her dreams of being a physical education teacher before the realities of life got in the way. During the time that I taped this woman, the young girl she had once been shone through, softening her features. I had succeeded in capturing the vulnerability and the hopes she once had had and that now lay below the surface of her physical appearance.

Kids and Props

Props can be an extremely valuable tool when working with kids. Their attention is easily diverted, so if you can keep their attention with a prop that they can play with, you have the best chance of

Figure 6–6 A bowling alley is an interesting location in which to shoot children with your vidcam. Kids naturally take to bowling and find the balls and pins fascinating. For the videographer, the ball can be used as an interesting prop, as in the picture shown here. Make sure to set the color temperature selector switch on your vidcam to indoor lighting when working in any indoor location. (*Photo by Curtis Southard, courtesy of Kodak.*)

shooting a good, candid tape. Often, the best props are the simplest.

A videographer I know finds it a constant challenge to come up with new props for children to use. Three years ago he noticed a tree near his home; it bent in an unusual way to one side. On a whim, he had a young girl whom he was taping for her parents climb the tree. The girl loved it. She could climb to her heart's content, and all the while the videographer shooting.

Many parents like to shoot their children in class at school. Books, writing tablets, inkwells, fingerpaints, erasers, chalk, and those small cartons of milk we all loved as children are props you can utilize. (By the way, if you intend to tape your child at school, tell him beforehand that you want to tape his whole class. In this way, you are not singling him out for exclusive attention, which might annoy or embarrass him in front of his peers.) And while you are there, do a tape of the school pageant. Who could ask for a better shot of your child than to show him dressed as Marc Antony, with cardboard sword at the waist and oversized spear at the ready, nervously waiting in the wings to go on in the school production of *Julius Caesar?*

Shooting a child at camp offers some novel prop possibilities. A child in a rowboat can radiate confidence and excitement. Or capture the zest for living each day anew that kids express, especially in outdoor surroundings, by following along on a hike. Be sure to enter into the experience of the place. It may help to remember what it was like to be a kid in the same situation.

Other Props

Musical instruments can also be used as props when taping. For instance, shoot a budding Al Hirt playing his trumpet at band practice. Shoot at the rehearsal, so you will have more time to set up than at the actual concert.

If there is a young chemist in your family, shoot him using the portable chemistry set he got as a Christmas gift. Ask him to describe his latest experiment as he recreates it for your camera. Let him hold a testtube, and you may see a Jonas Salk of the future through the viewfinder.

If a child or teenager engages in a particular sport, shoot him playing the sport or in his room at home where he has his pennants and trophies. He can hold a Louisville Slugger, football, or whatever

Figure 6–7 The use of props is a great way to get children to relax on camera. This child was stiff at first, but, when she began playing with a scissors and paper at the videographer's suggestion, she relaxed. Props will bring an added dimension to anything or any subject that you shoot. (*Photo courtesy of Kodak.*)

equipment his sport requires him to use. Remember, props can function as something to occupy the hands while the shot is being taken, as well as a physical and emotional extension of the individual's personality.

Interaction with Pets

Pets are an intimate facet of many families and are often thought of as family members. For the videographer, they provide a unique shooting prop.

Kids and animals are emotional, volatile beings who love each other. Rarely will you get a stiff shot when you tape them together. A tape of a child playing with his dog or of another child holding a purring cat in his lap gives the viewer warm, intimate feelings.

A sense of play and curiosity is natural to both pets and children. The spontaneity of their play should not be passed up in any taping session.

MAKEUP

You might want to consider using makeup on the faces of the people you are shooting. However, the complexion of a child is soft, rosy, and clean; it does not need to be augmented by artificial means. For most tapes of adults (excluding those done of professional models), a bit of pancake or rouge applied to the oily parts of the face will take the sheen off and give the skin a smooth, even appearance. Women might want to wear eye shadow, mascara, and lipstick, which is fine as long as it is not used in excess. On a tape where the subject is wearing too much makeup you wind up with a makeup advertisement and not a video portrait of the person.

CLOTHES

Remember that clothes are also an extension of the individual's personality. Generally, someone who favors flamboyant dress will tend to have an outgoing personality, while the person who dresses conservatively will tend to be more introspective.

Adults should wear color-coordinated clothes they feel comfortable in when you are taping them. If the subject is wearing a particularly outlandish or dull outfit, gently point out that you feel another outfit would be more appropriate. If they allow you and you have the time, help them pick it out.

When a taping session has been planned for children, they often get new clothes. Because parents buy clothes big enough to keep pace with the child's growth, the child may look lost in the new outfit. Make sure children are taped in clothes they feel comfortable in and that fit them well.

Figure 6–8 When shooting an older person, try to use your ability as a videographer to frame your shot so you get a real feeling for the subject, like photographer Curt Foster did here. The lines on a person's face tell the story of his life. (*Photo courtesy of Kodak.*)

HAIRSTYLE

While the child's hair should be neatly trimmed before a taping session, let the stiffness wear off for a couple of days before shooting. In this way, you can take time to get to know and establish a rapport with him, and he will not associate a taping situation with the discomfort of a new haircut. For adults, hair should be styled to frame and accentuate the face, not to detract from it. Nothing is more distracting from a video portrait than an outlandish hairstyle.

THE TAPE ITSELF

Self-esteem and ostentatiousness are only surface motives that conceal far deeper urges to have videotapes shot of oneself. Whether it be the garbageman who lives down the street or the corporation president in his Park Avenue mansion, that deeper urge is a person's desire to have an image of him or herself survive. In this respect, we are no different from our ancestors. The Greek girl who, centuries ago, traced the outline of her departing lover on a wall was inscribing the same sentiment that we, in our day, invoke with the aid of the vidcam and the VCR.

7

The Tuner/Timer

Now that your tape is shot, the time has come to play it back on your television set. Of course, you can play it back on the TV monitor of the electronic viewfinder, but squinting at the small screen can be tiring, and besides, the electronic viewfinder, as previously stated, can only reproduce the scene in black and white. It is much more enjoyable to watch a scene that was shot in color played back in color on a television set in your home.

You have two other choices for playback. First, you can connect the portable videocassette recorder to your television set directly and play the tape back in that way. We covered how to do this in Chapter 2. The second choice is to attach a tuner/timer to the portable VCR and do your playback in that way. Let's concentrate for now on the tuner/timer.

While our discussion so far has centered on shooting with your vidcam, many of you will want to purchase a separate tuner/timer to record directly off your television set.

By the way, for those of you doing your play back with a home deck, it has a tuner/timer built into it that functions the same way as the separate tuner/timer.

CONNECTIONS

In order to play back tapes on your VCR in conjunction with the tuner/timer, you have to make some connections between the tuner/timer and the VCR and between the tuner/timer and your television set. Here is how to make those connections.

If you look at the back of your tuner/timer, you will see that there are a number of cables sticking out of it.

The VTR connector cord, which is the thickest cable, should be connected to the tuner/adapter jack of the videocassette recorder. The RF connector cord, the next thickest cable, is connected to the RF out jack of the portable videocassette recorder.

There is also a remote control cord, which connects up to the remote control jack of the VCR. It provides remote control of the

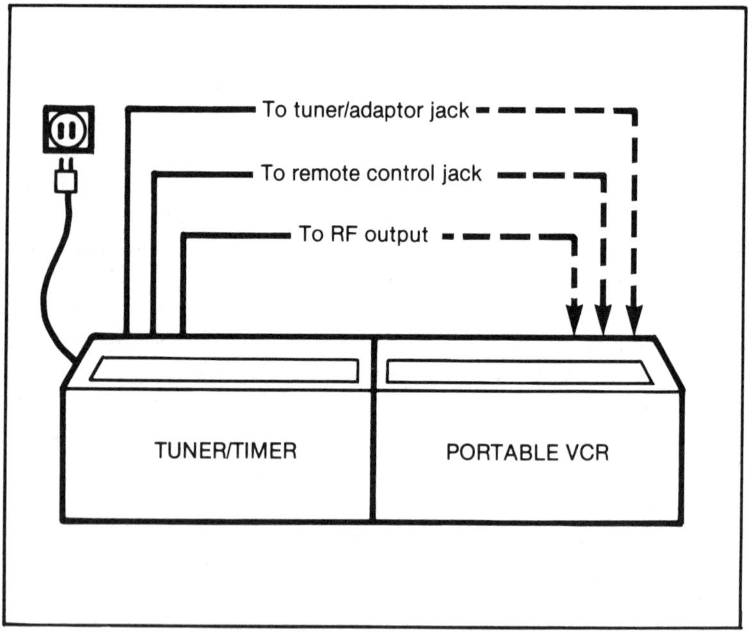

Figure 7–1 Connecting the portable VCR to the tuner/timer (front view). (*Adapted from diagram provided courtesy of Minolta.*)

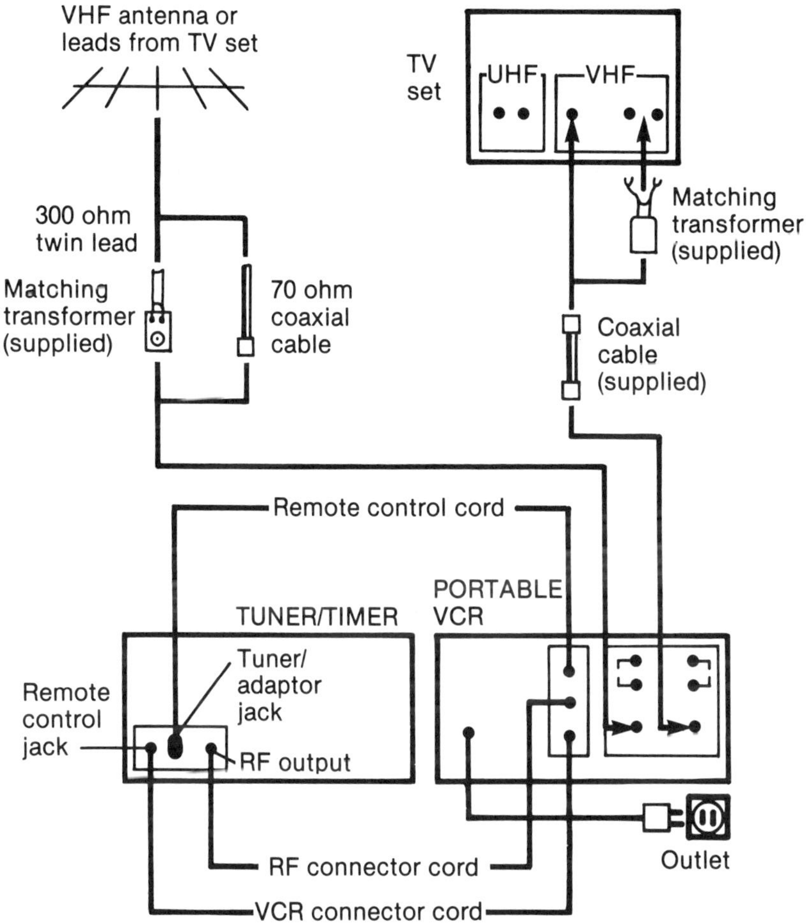

Figure 7–2 Connecting the portable VCR to the tuner/timer (rear view). (*Adapted from diagram provided courtesy of Minolta.*)

tuner/timer and the videocassette recorder when used with the remote control unit that comes with many tuner/timers.

Once you have made these connections, the next step is to wire the tuner/timer to the television set directly. To do so, use the 75-ohm coaxial cable that comes with the tuner/timer. Plug one end of it into the VHF antenna input terminal on the back of the tuner/timer, and then connect the other end to the matching transformer that comes with the unit. The matching transformer has on its end two prongs, which are then attached to the VHF terminals on the back of your television set. Then, take the other matching transformer (the one without the terminals) and plug it into the VHF antenna output terminal on the back of the tuner/timer. Attach the VHF terminals from your television set to the two screws at the bottom of this matching transformer. If you now plug in the AC power cord, which you will also find in the back of the tuner/timer, to any 120-volt AC household outlet, you will be all set to play back your tapes.

For those of you who have cable television, the 75-ohm coaxial cable should be plugged directly into the cable plug. You will not need the matching transformer, the one with the prongs.

RF CONVERTER

As mentioned in chapter 2, in order to have your VCR play back on your television set, you must set the RF converter on the back of the VCR to either channel 3 or channel 4, whichever is unused in your geographic area. The RF converter will change the video and audio signals from the tuner or tape playback to standard television broadcast signal.

THE PROGRAM TIMER

As soon as you plug the tuner/timer into your household outlet (or when a power failure continues for 60 minutes or more), the timer display on the front of the tuner/timer flashes 12:00 A.M. The brightness of the timer display can be altered with the dimmer button located on the control panel built into the fronts of most tuner/timers.

Now, to set the correct time, press and hold the clock button on the control panel. While you are holding this button, press the button for the day of the week — e.g., the button that says *TUE* (for Tuesday). The clock is now set to Tuesday. Still depressing the clock button, press the A.M./P.M. button, once for A.M. or twice for P.M.

To set the hours and minutes, while still depressing the clock button, press the numbered buttons in the following sequence: 1, then 1, then the Hr/Min button, then 5, then 2, and then the Hr/Min button again. Release the clock button. The time has just been set to Tuesday, 11:52 A.M. By setting the tuner/timer's clock to the accurate time, you can then record from your television set at the precise hour a television show comes on, so you do not miss a single minute of it.

TUNING

One of the functions of the tuner/timer is to assume control of the channel selection and tuning of your television set when the VCR is on. With both VCR and tuner/timer on, turn your television's channel selector switch to channel 3 or channel 4. This should correspond to the channel you have selected with the RF converter previously.

Set the tuner's output select switch to the VTR position. Now, by depressing any of the channel-select buttons on the front of the tuner/timer, you will instantly be switching channels. You will know what channel you are on because the number of the channel will be displayed and lit up in a little window below each of the channel selector switches.

Fine so far, but the picture might need some fine tuning. *Do not* reach for the television's fine tuner, because it will not work while the tuner/timer is in operation. Instead, open the fine-tuning access door, which is usually located on top of the tuner/timer. Find the fine-tuning knob that corresponds to the channel you have on and turn it until you get the best picture possible on the television set. Now go from channel to channel, switching with the tuner/timer. At each channel, stop and do your fine tuning until all the channels on your television set have been fine-tuned.

The videocassette recorder is now ready to begin recording from your television set or playing back a previously recorded tape. All you have to do, in either case, is insert the tape and you are ready to go.

SIMPLE PLAYBACK

Regardless of whether you have made a recording from your television set, bought a cassette of that fine B movie classic, *Wrestling Women versus the Aztec Mummy*, or just want to play back a tape shot last week of your child's school pageant, here is how to play back your tape.

Insert the videocassette into the VCR and turn it on. Depress the output selector switch on the tuner/timer to VTR and turn the television set manually to channel 3. Press the play button on the videocassette recorder. After a moment, the prerecorded tape will appear on channel 3, assuming, of course, that you have turned on the television set. By the way, the playback speed will automatically correspond with the recorded speed on many machines. Otherwise, you will have to set the playback speed manually to correspond with the recording speed.

After you have finished viewing the tape, press the stop button on the VCR, and the machine will automatically stop. If you have more unused tape on the videocassette, you may continue recording from that point onward. However, if you no longer want to look at the show you have recorded from your television set or the tape you shot in the field, simply press the VCR's rewind button. You can now record over the old tape, if you wish, unless the tapes were rented from a store. The record safety tab on the bottom of such cassettes is missing to stop you from recording over the rented tape.

RECORDING AND VIEWING THE SAME PROGRAM

To both view and record any program you are watching, insert the cassette into the VCR and select the proper tape speed. Next, select the channel you want to record on the tuner/timer. Finally, press the record button on the VCR and, while holding it down, also press the play button. The machine will record the show you are watching on TV.

If you want to omit commercials, press the pause button on the VCR when one begins, wait until the commercial is over, and then press the pause button again to continue recording. When the show is over, press the stop button and rewind.

RECORDING AND VIEWING DIFFERENT PROGRAMS

Back in the late 1960s, New York's channel 4 was broadcasting the original run of *Star Trek* while channel 7 countered with *Judd for the Defense* on Friday nights. It was a continual source of friction in our one-television family. Usually, my sister and I opted for the adventures of the crew of the *USS Enterprise,* while my father and mother wanted to watch the story of Clinton Judd, an attorney who flew around the country defending the underdog. We could never seem to resolve our conflict, so one week we watched *Star Trek* and the next week we watched *Judd for the Defense.*

If we had had a VCR and tuner/timer, there would not have been a problem. We could have inserted our cassette into the VCR and set the VCR's tape speed. Next, we would have selected the channel to be recorded by depressing the channel-selector button on the tuner/timer. On the week my sister and I had our way, we would have recorded channel 7. On the week my parents had their way, we would have recorded channel 4. Then we would have set the output selector switch of the tuner/timer to TV, and of course, the television set would have been turned manually to the channel we wanted to watch while the other channel was being recorded.

Once all this was done, we would have pressed the record button, holding it down while simultaneously pressing the play button. The machine would then have recorded the program on the channel selected by the tuner/timer while we watched the program selected by the television set.

When we were finished recording, we would have depressed the stop button and then rewound the tape. The recorded program would be on videotape for viewing at our leisure, and all family arguments would have been averted.

TIMER RECORDING

Timer recording really involves programming your tuner/timer to turn on the videocassette recorder at a preset time, record the show you want to watch, and then turn the VCR off when the show is over — all without any intervention from you, since presumably you will be away from your television set. While programming controls differ from model to model and manufacturer to manufacturer, they are all basically alike. The first thing to do to record a show is to open the control panel on the front of the tuner/timer. Press the program button and then the number 1. This tells the machine that you are about to program in information to record one show.

Next, select the day the VCR is to start. For *Hill Street Blues*, for instance, the day is Thursday, so press the button that says *THU.* Then press the A.M./P.M. button twice. This sets the machine's clock to P.M. For a show during the morning, set the clock to A.M. by pressing the button once.

Select the time the VCR is to start. *Hill Street Blues* begins at 10 P.M., so press the buttons on the control panel of the tuner/timer in the following sequence: 1, then 0, then the HR/MIN button once, then 0 twice. This tells the machine to begin recording at 10 P.M.

To program the machine to turn off at the conclusion of the show, press the A.M./P.M. button twice, then 1 twice, then the HR/MIN button once, then 0 twice. Press the channel-selector button on the tuner/timer to let the machine know what channel is to be recorded.

Now the tuner/timer's digital display will light up, telling you that it will tune the videocassette recorder to the proper channel on Thursday at 10 P.M. and will turn it off at 11 P.M. When you arrive home from wherever you have been, the recording will already have been made — but only if you have remembered to press the timer button on the tuner/timer. This activates the programming information in the machine so that it is turned on to record at the time you have specified.

The tuner/timer allows us the luxury of unattended recording at any hour of the day or night. It may be convenient, for example, to record a late movie while you are asleep — or perhaps the 11 o'clock news for playback the following morning while you are getting ready for work.

Some tuner/timers are also capable of recording on a specified day at the same time each week, recording at the same time every day of the week, and recording on any day of the next week or the week after next. As you can see, this type of programming can get more detailed, and controls can vary from machine to machine. If you want to do any of this sort of programming, check the instruction manual that comes with your tuner/timer for complete details.

BATTERY CHARGING

One of the added benefits of having a tuner/timer is that you do not have to spend additional money on an AC charger/adapter that is used to charge up the battery that comes with the videocassette recorder. The tuner/timer can take care of the charging for you.

To do this, make sure the battery has been inserted in the back of the VCR and plugged into the proper jack. Next, make sure the power switch of the VCR is off. Press the charge switch located on the front of the tuner/timer. Now the battery will begin charging.

When the charging starts, the charge indicator on the tuner/timer will light up in red. When the charging is completed, the charge indicator will go off automatically and the charge switch will be disengaged. I should mention, though, that if you should accidentally turn on the VCR, the charging will immediately stop. So, when doing your charging, be sure that there is no power going to the videocassette recorder.

You can also charge an auxiliary battery pack using the tuner/timer. Just connect the battery pack to the auxiliary battery charge terminal of the tuner/timer, which is located in the rear, and press the charge switch. As before, the charging indicator will glow while the battery pack is being charged and will go off when the charging is complete.

When charging an auxiliary powerpack, the power switch of the VCR can be on or off. In addition, the video tuner and the VCR do not have to be connected when charging a battery at the auxiliary battery jack. The tuner/timer, though, must always be connected to an AC outlet to perform any of the functions mentioned in this chapter.

REMOTE CONTROL

The remote control unit that comes with many tuner/timers will not enable you to operate a robot, but it will operate your tuner/timer and VCR without your having to touch the buttons of either.

Operation

In order to operate the remote control unit, make sure that during the connection of the VCR to the tuner/timer you remembered to plug in the remote control cord that goes from the tuner/timer into the VCR. Once this is done, open the back of the remote control unit, insert four 1.5-volt AA batteries, and then close the battery compartment. Now it is ready for operation.

The remote control unit works by infrared signal. There is a small panel on the front of the tuner/timer; it is called the *infrared receiving section.* This panel receives the infrared signal emitted by the remote control unit when one of its control buttons is pressed. The infrared receiving section then commands the tuner/timer or the VCR to perform one of its functions, corresponding to the button pushed on the remote control unit.

Functions

In addition to the functions of the VCR and the tuner/timer already described, there are a few others that may not be on the VCR and tuner/timer, but which the remote control unit is capable of performing. All the controls on the VCR and tuner/timer can be operated from a distance using the remote control unit. You have to be in fairly close proximity to the unit — no more than approximately 10 to 15 feet — and if something gets in the way of the infrared signal, such as a piece of furniture, the signal will not be received.

Frame Advance. The Frame Advance capability of the remote control unit operates only when a tape has been recorded in the slower SP or EP modes. To advance the picture a few frames at a time during playback, first press the pause button on the remote unit and then the frame-advance button. The picture will be advanced a few frames at a time each time the latter button is pressed. The frames will be advanced continuously when this button is held down.

Frame advance is especially useful if you have taped, say, an athletic event, and you want to see the step-by-step execution of a difficult athletic activity or play. It is also good for viewing close plays, such as the case with slow-motion instant replay in broadcast television.

To release the frame-advance mode, press the pause button again and the VCR goes back to regular playback speed. Please note that there will be a certain amount of "noise" in the frame-advance mode.

Slow Advance. The slow button on the remote control unit plays back your tapes in slow motion, but it only operates at certain speeds, such as in the SP or EP modes. To play back your tapes in slow motion, press the slow button. You can then use variable slow control buttons to vary the playback speed from one-fifth normal speed to stopped action in the EP mode, or from one-tenth normal speed to stopped action in the SP mode. When you want to go back to normal speed, simply press the play button and the machine will begin playing the tape back at normal speed.

The playback picture will have some "noise" in the slow mode. Press the forward or reverse slow tracking buttons to minimize the noise bands that may appear when using the slow, still, or frame-advance buttons. This should enable you to move the noise band to either the top or the bottom of the picture.

Visual search. This is probably the most important VCR feature, in addition to extended programming capabilities. And the desire for it is understandable. The visual search button allows you to view the picture on a recorded tape at high speed in both the forward and reverse directions. It can only be operated at some speeds, such as in the SP and EP tape modes.

To use the visual-search capability of your remote control unit, press the play button to play back your tape. Now press and hold down the forward visual-search button. If you then look at the TV screen, you will see the picture racing by. If you pass the point of interest to you, just depress the visual-search button that operates in the reverse direction. When you get to the point where you want to go back to regular playback speed, take your finger off the visual-search button. After a moment, regular playback speed will resume.

Tapes recorded in the SP and EP modes will be played back at four times the ordinary speed (SP) or at twelve times the ordinary speed (EP) when the visual-search button is employed.

Special instructions. While an obstacle between the remote unit and the VCR will probably inhibit the remote unit's function, infrared rays can bounce off nearby objects and so might hit the infrared section of the tuner/timer and make remote operation possible. Such operation may not be consistent, however, so do not depend on reflected rays to operate your machine. Have a clear path between the remote unit and the tuner/timer.

As the batteries of the remote unit are used up, the operating distance is reduced. When operation is not possible, or when the operating distance is sharply reduced, replace the batteries.

Finally, when using the remote unit to operate the various functions of your VCR and tuner/timer, a slight delay in reaction time is normal. When depressing the function buttons on the remote unit, make certain that each function has been activated before releasing the buttons and selecting the next operation.

SELECTING A TUNER/TIMER

Now that we have considered all the functions and connections of the tuner/timer, including the operation of the remote control unit, let's take a look at some of the many tuner/timers you may want to consider purchasing.

Sony

Sony's tuner/timer is made to function with any of their portable videocassette recorders and vidcams. The newest tuner/timer the company has out is the Model TT-200, which is the same size as their Model SL-2000. The TT-200 has a built-in 2-week, four-event timer. It covers the complete VHF/UHF broadcast range. There is also a wireless remote control unit that is supplied with the TT-200 for use with the Betamax recorder and tuner/timer.

This versatile remote controller operates a number of recorder and tuner/timer functions, including power on/off, channel up/down selection, program selection (VTR/TV), recording, play, fast forward, reverse, Betascan, stop, playback, pause, freeze, and Videola (normal and slow motion forward and reverse).

Quasar

Quasar's tuner/timer was introduced in the middle of 1983. It is the VH5125UW tuner/timer. This is a front-loading model that is electronically tuned. It has two-way picture-search capability, freeze-frame control, and other Synchro Touch Controls, as well as four-function remote control. Programmability is limited to 1 day and one program.

Canon

The Canon tuner/timer operates with the company's video recorder and video camera. The Canon VT-10A provides the flexibility and features needed for most types of home recording. Fourteen feather-touch control buttons (for channels 2 to 83) let you precisely tune

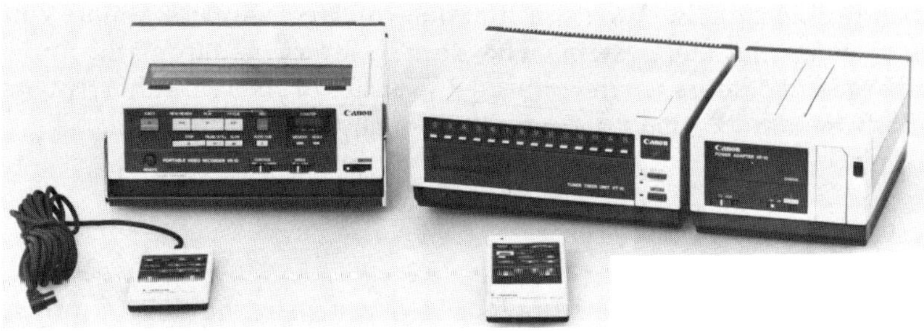

Figure 7–3 With the Canon tuner/timer, you can tune in any television channel in your area and then tape off it, whether you are home or away. The power adapter allows you to run your VCR on simple household current and to charge your VCR batteries. The VCR will also run if it is attached to the tuner/timer.

any on-air station. The tuner is also equipped with a wireless remote control (Model WL-10A), and this unit is cable-ready.

The VT-10A is programmable up for to four programs in a 2-week period. Repeat recording is also possible. The timer display consists of a quartz-controlled oscillator with a 12-hour clock system and 12-hour digital display. During regular operation, the tuner/timer also acts as a charger for the built-in recorder battery.

General Electric

GE is producing a large line of video equipment, including two tuner/timers. (It is interesting to note that whereas the camera companies who are making video equipment have only one tuner/timer to go with their recorders and cameras, the video companies provide more of a choice.)

The first GE tuner/timer is the Model 1CVT605. It has a 24-hour, single-event timer. There is also a fourteen-position, eighty-two-channel electronic tuner and twelve-function wired remote control capability.

The Model 1CVT615 tuner/timer has a four-event, 14-day programmable function. There is also 1-hour memory protection in case of a power failure. The unit features a fourteen-position, 105-channel tuner that is capable of receiving unscrambled midband and superband cable signals. A sixteen-function infrared wireless remote control completes the system. The deck's battery is automatically recharged in 5 hours when connected to either of the two tuner/timers. Both models also have a simple one-touch recording system.

Minolta

An integral facet of the Minolta video system is the T-700S tuner/timer. This unit, when combined with the Minolta V-700S recorder, permits unmanned recording of television programs for up to 3 weeks. The tuner also charges the batteries in the Minolta recorder, eliminating the need for the adapter/charger.

The T-700S tuner/timer is an easy unit to operate. In addition to tuning in television programs for recording while they are being viewed, the unit can be programmed to record shows on a channel other than the one being watched.

Changing channels or using the recorder's special effects — single-frame advance, slow motion, freeze frame, forward and reverse speed search — are all done with the new seventeen-function Minolta RC-1S remote control unit, which is supplied with the tuner/timer. It is wireless, so there are no inconvenient wires to clutter up the room; its infrared signal provides spontaneous control for all tuner and recorder functions. In addition, the tuner/timer comes with various extension cables to facilitate connections.

Zenith

The Zenith VRT9850 tuner/timer has a fourteen-position electronic voltage synthesizer with a signal-seeking control. The timer can record as many as four different channels over a 14-day period.

An infrared full-function wireless remote unit puts you in control of all tuner/timer and recorder functions: tape/TV, record, play, fast forward, reverse, stop, pause, special effects, power on/off, and channel change from anywhere in the room.

This unit also provides power to operate the Zenith VCR on AC household current and to recharge the nickel-cadmium battery pack that comes with the Zenith VCR. The VRT9850 can also recharge an auxiliary battery pack.

JVC

Styled to match the JVC HR2650U recorder, the TU-26U tuner/timer allows for unattended recording and battery charging for the JVC recorder's battery pack; an AC power adapter is also provided.

This cable-ready unit receives a total of 105 channels, including UHF, VHF, midband, and superband frequencies. Fourteen channels can be preset for convenient sweep-tuning selection.

A 14-day, eight-event programmable timer incorporating a microcomputer provides the opportunity for away-from-home recording and time-shift viewing. A channel-lock system prevents accidental changing of channels during recording. This control is automatically engaged during timer recording to ensure that the preset stations are the ones recorded.

The time of day and all necessary recording information are indicated by a bright fluorescent digital display on the front of the unit, which includes a 12-hour clock and A.M./P.M. indicator.

Finally, the TU-26U recharges the JVC recorder's battery pack in 90 minutes and converts regular AC household current to DC power for the HR-2650U recorder.

Sanyo

Sanyo's tuner/timer is the Model VTT 481; it operates in the Beta format. It lets you program five events up to 14 days in advance, so you never have to miss your favorite program, even when you are on vacation.

The built-in AC adapter and battery charger supplies power for the Sanyo VPR 4800 videocassette recorder. A fluorescent display on the front of the unit provides continuous time of day readout. It also serves as a program timer display.

Other features include LED indicators for power on, channel-selection capability, clock and timer set buttons, automatic fine tuning, TV/VCR selector, audio/video input and output jacks, and jacks for antenna inputs and outputs.

Sylvania

The VC-4526 tuner/timer has a special memory circuit that permits it to be programmed to record up to four programs in a 14-day period.

Figure 7–4 The Sylvania tuner/timer, No. 4526, operates with the Sylvania portable VCR. The tuner/timer is pictured here on the right.

This electronic pushbutton 105-channel tuner allows you to watch one program while recording another. It comes with a sixteen-function infrared remote control unit.

Philco

Philco's tuner/timer works with the company's VHS cameras and recorder. It has the capability of being programmed 14 days in advance at three recording speeds: SP, LP, and EP. There is a fourteen-position electronic VHF/UHF control with automatic fine-tuning capability.

Other controls on the tuner/timer include variable-speed slow motion, video dub, video cue/review, still framing, tuner backup, midband/superband, digital clock and built-in timer, quick battery charge, automatic stop and rewind, manual pause button, audio dubbing button, and video and audio input and output jacks. A carrying strap and a battery for the recorder unit are included when you purchase this tuner/timer.

Magnavox

Magnavox calls its tuner/timer a "control center"—Model VR-8382BK. With this unit you have audio and video dub capabilities, as well as fourteen channels you can program automatically. Moreover, this unit is capable of charging up the battery that comes with the Magnavox recorder in under 3 hours.

Figure 7–5 The Magnavox "control center" is pictured here on the right. On the left is the portable VCR that it works with.

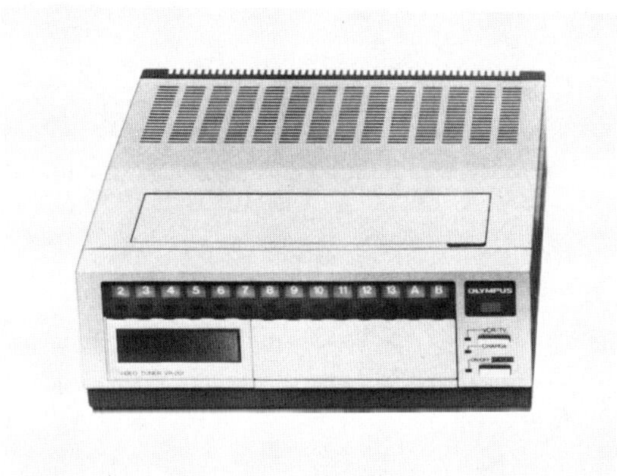

Figure 7–6 The Olympus tuner/timer has a fourteen-channel, soft-touch electronic tuner. It is quartz-controlled and can be powered by any 120-volt AC outlet.

Olympus

The Olympus Model VR-201 programmable video tuner/timer permits taping of any TV show for later viewing and can be programmed to tape up to four programs during a 2-week period. With the VR-201, you can also tape one program while viewing another. A compact twelve-function wireless remote control unit is supplied without additional cost.

8

Making Money with Video

In today's economy it is sometimes useful to have a second source of income. A few weeks ago, as I was driving home late from New York City, my headlights flashed across a sign on a lamppost that said: Video weddings shot. Call Vinnie at Northside 7777. So I called him, just to find out what kind of business he was running.

It turned out that Vinnie was a sanitation worker who had bought a video outfit to shoot his family. Over a period of 2 years he had gotten proficient enough that his friends had urged him to turn professional. Last summer, he finally decided to try his hand at shooting weddings with his vidcam. He found that he could make a good supplemental income. He says that if business keeps up the way it is, he may consider shooting full time.

THE WEDDING MARKET

One of the most accessible markets for making money with your video equipment is the wedding market. For years the impregnable bastion of the still photographer, weddings are now being shot more and more by videographers.

The generation of people being married today is the first to be raised on the televised image. The TV has been the babysitter to kids when their parents were not home, their companion as single adults, and now, the medium for recording the most important of life's events, among them, their weddings.

Rudy Pollack, the owner of Dino Photography Studio in Elmhurst, New York has been shooting weddings on video for 12 years. He has some definite suggestions on breaking into the wedding market.

"You should contact your local photography studio, one that you know is doing lots of weddings," Pollack says. "See if you can subcontract the work from the studio. If you can, and you should be able to because video weddings are now popular, I'm sure that you can get plenty of work."

Pollack cautions against going to a florist or a caterer for referrals, because a bride and groom will not book a wedding through these parties.

"The bride and groom want to go to a place that specializes in the needs they have," Pollack explains. "They won't go to a florist for video, they won't go to a musician for video, but they will come to a photographer to book a video wedding. Photography and video have a natural association."

By taking the time to canvas the wedding studios in your area, you should be able to come up with one that does a heavy volume of business and is willing to enter into a subcontractual agreement with you. Before approaching the studio, though, it is a good idea to shoot the wedding of a friend or relative. Then use that tape as your show piece. A video project done in school can also be used to display your video talents.

Equipment

Once you have entered into a signed agreement, one that specifies what you will be paid and what the studio's percentage is, the next step is deciding what equipment you will need for the actual shoot. On this point, Pollack is particularly adamant.

"The absolute, most important thing is to have backup equipment. If the vidcam fails, you can't turn around and tell the bride and groom, 'Look, my camera failed.' They don't want to hear it. You'll be sued for not living up to your end of the agreement."

"You have to be prepared to go to the wedding with two of everything, just like the still photographer does. This means two cameras, two videocassette recorders, two battery packs, two mikes, etc."

Speaking of mikes, consider using wireless ones whenever possible. If you are trying to pick up the bride and groom speaking while the band plays and you are using the in-camera mike, all you will pick up is the sound from the band.

Lighting

Lighting the wedding is extremely important. Pollack recommends using quartz lighting, as opposed to the cheaper photofloods.

"You should know how to control your lighting. You just can't do a professional video job without supplemental lighting, because in the average catering hall you can't see your hand in front of your face."

He explains that bounce lighting cuts down on glare, providing a soft, even illumination. Bounce light also makes an excellent fill light, filling in the shadowy portions of the subject's face.

The Affair

Once you have decided on the equipment you are going to use and you have two of everything, your next consideration is the specific shooting scenario. What shots do the bride and groom expect you to get? Are there any special family traditions that will take place at the wedding that you should be ready to shoot? In short, what is it necessary to get on tape so that you have an accurate representation of the affair?

Like the ubiquitous wedding photographer, divide the wedding into three sections: the formals, the ceremony, and the reception.

During the formals, take group shots of all the wedding principals: bride, groom, parents, grandparents, sisters, brothers, etc. You can shoot these scenes at the church or synagogue, or if you have time, go over to a local park. There the beautiful location will reflect the beauty of the upcoming event.

Next comes the ceremony. You *must* get permission in advance to shoot the ceremony from the priest or rabbi. Many houses of wor-

Figure 8–1 The wedding reception is a great time to get candid shots of the bride, groom and family. These are once in a lifetime moments, so perfect your video shooting technique before going out on a job for hire. You may get the opportunity to tape over bad shots, but you will not get the opportunity to do the *same* job a second time. (*Photographs by Ken Sklute for Dino Custom Color Photography, Elmhurst, NY.*)

ship will object to having lights set up for the ceremony, but they will allow you to shoot with available light.

Be sure to check out the church or synagogue in advance, so you know where the best angles are to get the bridal procession and the wedding vows on tape. Then, it is a simple matter to change camera angles and lens focal lengths to add some video spice to the proceedings.

It is during the reception that Pollack particularly suggests the use of good, varied shooting technique. "Emotions are very important," he reveals. "For instance, the bride and her father are dancing.

Figure 8–2 Another way to make money with your vidcam is to shoot the valuables of your neighbors on tape for insurance purposes. Insurance companies require some sort of proof before they will pay on robbery claims. By shooting valuables, you provide them with visual proof that is irrefutable. (*Photo courtesy of Kodak.*)

You don't open with a wide shot of the entire room. You come in tight with a close-up and you get the emotion and the crying. That's what makes video — the emotion."

Pollack advocates learning in-camera editing techniques because a wedding videotape should be 45 to 60 minutes long, not longer. You have to make sure each shot counts.

"Make sure, too, to mix up your shots during the reception," he says. "While a close-up lends intimacy, establishing shots of the reception hall, medium shots of the couples dancing, singing, and eating, and then close-ups of the bride and groom proudly taking in the whole scene will give a more complete record of the happy day. Remember, too, to get the best man's toast, the formal introduction of the newlyweds to the crowd, the cake cutting, and the tossing of the bridal bouquet. These are moments that will never be repeated. You will have only one chance to capture them, and you have to make it count."

This is a good time to remind you to get insurance before shooting the wedding. You have to have it in case someone trips on one of the connecting cables. In New York City, for instance, many of the catering halls will not allow you to shoot unless you show them a certificate of insurance. So get the insurance and carry proof with you when you go out to shoot.

Finally, since you are subcontracting, you will be giving the material to the photography studio, which will handle the delivery to the bride and groom. If you charge the photography studio $250 for your services, that is a good, reasonable price. The photographer will probably mark the price up to around $500, but he is the one providing you with the contacts and, therefore, the business.

After you have been shooting professionally for a while and have gained some expertise, you can begin charging more for your services. As you get your own clients, your fee will continue to be enhanced in proportion to your ability to deliver a high-quality tape.

LITTLE LEAGUE GAMES

Another type of tape you might consider shooting professionally is a tape of a Little League baseball game. Personally, I was never a very good baseball player. Still, if someone had offered my father a tape of my Little League "career," like any proud parent he would have jumped at the chance to purchase it. Especially for children, sports are a big part of their leisure-time activity. Consequently, many children play baseball or some other organized sport. Why not capture their "careers" on tape and then sell the tape to their parents?

Getting the Contract

The Little League season begins in April and ends in June. Summer leagues extend through the summer. If you want to crack this market, then it means giving up sitting in front of your television set in the spring and summer.

In the fall, teams begin forming for the upcoming season. That is the time to approach the executive board of your local Little League to land a contract, which is a necessity before doing any taping.

Elliot Lipson, past president and current member of the board

Figure 8–3 Shooting Little Leaguers is a great way to make extra money. Someday, you may be able to say that you shot a potential big leaguer. (*Photo by Elliot Lipson.*)

of directors of the New York-based West Brighton Little League, advises video shooters to come up with a concrete proposal about what they want to tape and when they want to do it.

"Once you've done that," Lipson says, "present the proposal to the board of directors of your local Little League. The board, which represents the league in all negotiations with outside parties, will want to know what their share of the take will be before they let a contract that will allow you to tape their players."

In other words, the league will want a payment in return for permission to tape their players. It might be 10% of your gross revenues, or it might depend on a barter system, where you agree to tape all team functions in return for the league's permission to tape their players. Whatever contract you let, be sure it is in your best interests and that it is in writing; then stick to it. As you build your business, your reputation will precede you.

The easiest way to sell your Little League taping service to the players and their parents is to adopt the practice used by professional team still photographers. What they do, and what you should consider, is to make up a flyer or advertisement. This paper explains your service and prices. Give it to the team manager, who will then distribute it to his players. The contract with the league's board of directors should specify the manager's assistance in this matter.

Once this is done, the players will take the information home to their parents, who will make the monetary decision. After a parent agrees to purchase the tape, you can schedule a shooting time that is agreeable to you, the parent, and the player.

Be sure to give parents a choice of purchasing two types of tapes. The first choice is a tape of the player during a game. The script for this video service might go something like this:

Individual tapes. You arrive at the Little League field before anyone else. You get your equipment in working order and take a few shots for practice. Pay particular attention to the various places on the sidelines you can shoot from.

Start your tape with an establishing shot of the field and the players and parents arriving. Cut to a close-up of Jimmy, your pint-sized subject for the afternoon. Catch the expression on his face. Kids have different reactions to playing ball: expectation, confidence, fear.

Have a variety of shots showing Jimmy warming up with his teammates. Then cut to a medium shot of the bleachers at the far end of the field filled with screaming parents reacting to a play. This will tell the viewer that the game has already begun.

In a number of shots, contrast Jimmy, who is intently watching the opposing pitcher's "stuff," with medium shots and close-ups of Jimmy at bat. If he gets a hit, pan with him as he rounds the bases. If he winds up coming home on a close play, you *must* be there to get him sliding in. Use a telephoto focal length, combined with a close-up, so the lens will be close enough to capture the action but not so close as to block the umpire's view.

If Jimmy grounds out, you should still follow him down the baseline, as you should if he strikes out. Expressions are important, so be sure to get enough close-ups to tell the emotional story.

Throughout the tape, intersperse shots of Jimmy's parents, who are sitting in the bleachers reacting to their child's play. Often parents will be surprised at their own rather flamboyant reactions.

The next to last shot could be Jimmy leaving the ball field and driving home with his parents. The last shot could be the now-deserted field, where so much action just took place and now is alive only in memories (and tape).

The highlight tape. The second choice of tape you can offer parents is a highlight tape. It is just what it purports to be — highlights of the team's play during the past season. To do a highlight tape, combine shots of the individual players whom you have been shooting all season and edit them down to a single tape. Any player you did not shoot in the first place is not going to want a highlight tape of his teammates' play, so your market is going to be the parents who have already purchased the individual tape as well.

Pricing

Elliot Lipson notes that an average Little League team has fourteen players and that there are four to ten teams in a league. So if you charge $75 for individual tapes and $50 for highlight tapes, and if you decide to shoot at least five teams, with half the players buying your service, your minimum gross billing will be $1750. Take off 10% for the league's share and you have just made $1585. That is enough to buy a new camera and VCR.

Try the Game Out

Before taping baseball; or any sport for that matter, it helps to try the game out. You cannot know what a game is all about until you try it. So if you are going to tape a Little League game, go out and hit a few with a friend, play a game of catch, and most important of all, remember what it was like when you were a kid. Remember standing in the outfield waiting for the ball to be hit to you and hoping that you would catch it so you'd be the hero and not the goat? It is this type of memory that will make you a better videographer.

REUNION TAPES

A nostalgia craze has swept the country, and many people are making money by exploiting people's memories. If everyone else can make money on nostalgia, why can't you? Everyone has fond memories of earlier times, and it is that feeling of yearning for them and the

A

B

C

Figure 8-4A, B, and C Bar mitzvahs and confirmations are also a good place to make money shooting video. If you decide to shoot one, divide it up in much the same way you would a wedding. First, shoot the formals, when the family gathers prior to the ceremony, as in Figure 8-4A. Then, shoot the ceremony, as seen in Figure 8-4B. Finally, after the ceremony is over, everyone gathers for the party afterwards, which is shown in Figure 8-4C. (*Photographs by Allan Berman.*)

desire to know what people are doing now, to see them once again in person, that attracts people to reunions. To sell a tape of a reunion to the participants, or to someone who could not be there but still wanted to be a part of it, is a natural business proposition.

Getting the Business

How to get the business is your first consideration. Most schools have alumni associations. Contact them and find out when and where their reunions are being scheduled. Remember, each high school and college class that graduated is going to have a reunion, which means the market of potential buyers will be quite large.

Another way of finding out where and when reunions will take place, especially in smaller towns, is by listening to the radio. As a

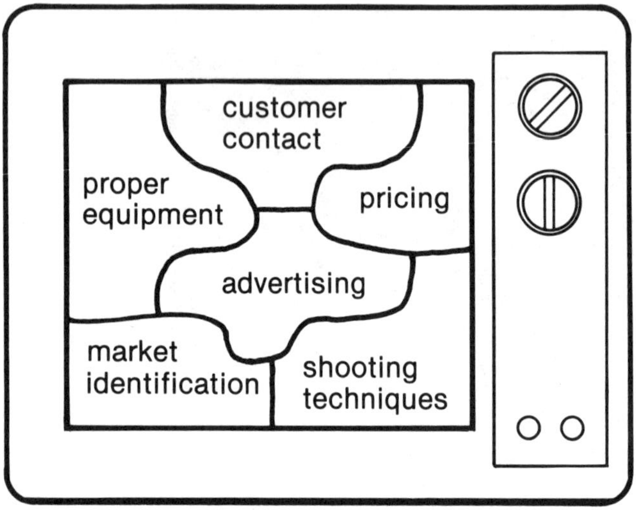

Figure 8–5 There are many factors you must consider if you plan on making money with your video equipment.

public service announcement, local radio stations frequently will mention scheduled reunions. It is a simple matter, once you have found out where and when the reunion is going to take place, to contact the alumni association and arrange to tape the event. Like the Little League situation, you will have to give them a portion of your revenues to get the contract, but it could be worth it, because hundreds of people attend these events.

Once the alumni association agrees to let you tape the event, get access to their mailing list. Make up postcards explaining to each former student that you are taping the reunion and ask if they would be interested in purchasing a tape. If the response is positive, it pays to go ahead with the taping. If the response is negative, forget that reunion and find another one, where the invitees are freer with their money.

Factors to Consider

Portability is going to be the biggest technical factor you should consider when taping a class reunion. Since you are going to be carrying the vidcam and VCR around for hours, the lighter the better.

Since the halls and gymnasiums where reunions take place tend

Figure 8–6 If you get really proficient with a vidcam and VCR, you may want to consider turning professional. In addition to the other ideas for marketing your video services mentioned in this chapter, you may want to consider shooting the news for your local TV station. Their video equipment, while of professional quality, operates in the same way as the amateur equipment described in this book. And, it is easier to break into shooting the news on the local level, because there is less competition and fewer union regulations than you will encounter on the national level. (*Photograph by Mara Bovsun.*)

to have rather dim lighting, take along some quartz lights on tripods that you can place strategically around the shooting area to provide you with enough light for taping.

Another point: *Don't* set your equipment down anyplace you are not, even if someone offers to watch it while you go to the bathroom. There is always someone who, in an alcoholic haze, will stumble over and break your equipment. Make sure you keep your equipment with you at all times, and do not let someone else use it.

The Shooting Scenario

What should you shoot during a class reunion? There was a film a few years ago called *Fraternity Row* that began with a pan across

Figure 8–7 Reunions are a marvelous place to shoot video for profit. Be sure to bring along an earphone, like the operator in the picture is wearing, so you can monitor the audio going onto the tape. (*Photo by Ken Sklute for Dino Custom Color Photography, Elmhurst, NY.*)

the pages of a yearbook, showing the faces of long-ago graduates. You could begin your tape the same way.

Ask the alumni association for a copy of the yearbook of the class you will be taping. Put it on a copystand, set up some photofloods, and pan slowly through the pages. As a narration, add some literary quotes you are fond of and that relate to getting older. It might be a passage from *The Great Gatsby* or *You Can't Go Home Again*. The important thing is to get into the nostalgic feeling of the event.

Fade from the yearbook to the faces of the people arriving in the gymnasium, community room, or hotel ballroom where the reunion is taking place. You *must* be there for the arrivals, for they will be the most important emotional moments of the night.

Figure 8–8 Shooting a model or an actress's composite tape is an excellent way to make extra money. You can shoot indoors, or on location as shown here. (*Model is Victoria D'Angelo and photograph is by the author.*)

Emotions will flow as freely as wine, and you have to capture those emotions on tape. Otherwise, you are wasting your time being there. You will be using a lot of close-ups and medium shots during this part of the proceedings, because you want to be as intimate as possible.

During the evening, as you move about the room, interview some of the people on tape. You will not be able to get everybody, of course, but it is amazing how one person's life parallels so many others. For instance, ask the dark-haired fellow in the corner what he thought he was going to do for a living when he was going to school and what he is doing now. What about Sheila, the star high school athlete? Why did she give up a promising athletic career, marry Joe, and raise children before she reached her potential? How did Jerry, the class idealist, become the class pragmatist?

At some point during the proceedings, have the master of ceremonies make an announcement that you are taking orders for copies of the tape. Set up a small table in the lobby or the rear of the room, where people can come over and fill out order forms as they leave or go to the bathroom. It will not hurt to have a friend come along to staff the table, so you can always be free for taping.

Pricing

If you charge $50 a tape, and twenty people at a reunion take advantage of the offer, you have made $1000 before expenses. However, since there will probably be many more people who want to buy the tape, the profits should be quite good, even after the alumni association collects its share.

After the reunion, check your acceptances against the reunion mailing list and do a mailing to those people who have not as yet committed to purchasing a tape.

AUDITION TAPES

John Schaub is a professional photographer based in New Hyde Park, New York, who in the past few years has branched out into shooting videotapes for money. At first he shot video weddings. Having had years of experience as a still photographer, he was able to take his knowledge and use it when he went out to shoot his wedding videotapes. Since John has his own photography studio, he was also able to do quite a bit of portfolio photography for models and actors.

Composite photographs are made up to show the model or actor in a variety of costumes and poses. The model or actor tries to use a different facial expression in each photograph to show his range. The shots are then combined on one photograph and sent to casting agents and advertising agencies in the hope of getting jobs for the model and actor. John Schaub realized, though, that videotape was the coming thing, and he decided to branch out even farther by shooting actors' composites on videotape.

Advertising

John's first step was to advertise his services in the two show business publications that have affordable advertising rates: *Show Business* and *Backstage*. He advertised that he would film actors, who could play their favorite scene on videotape; models, who could be shot in a variety of eye-catching outfits; and even comedians doing their material. He decided to charge $100 for a tape done in his studio, while a tape done on location would cost $400. The response was good.

Self-Promotion

What these videotapes do is effectively promote the actor, model, or comedian. The people who take advantage of this service are not established stars obviously, but are people who are looking for a break. Consequently, they may not have much money. They may say they will pay you for your tape after it is finished or after it has been sent around.

In such a situation, you would be very foolish to go along with such a proposal. Since people working in show business are sometimes, as the saying goes, here today and gone tomorrow, you must work out a pricing scale and payment plan that protects you.

Schaub found it useful to charge the client 50% of his fee in advance, in *cash*. The remaining amount of the fee was payable upon delivery of the videotape — once again, in *cash*. If you get to know the person you are shooting and eventually shoot them again, at that time you may trust them to pay by check, but until that time it should be cash only.

Technical Requirements

Schaub found that the technical requirements for shooting an audition tape were not that great. What he needed was his vidcam, a tripod, an extra battery pack, the VCR, and a couple of videotapes.

He wrote out in advance a rough script of what the client wanted him to shoot. If he was shooting a model, he noted what the different outfits she wanted to wear were and what kind of posing routine she wanted to go through. If he was shooting an actor doing a scene from a play, he planned how he would arrange his shots. At what point would he move in for his close-ups to get the expression on the actor's face and when would he change camera angles? For the comedian, he had to take into consideration how much of the person

Figure 8–9 If you tape a model during a modeling set-up, you will have to get access to a photographer's studio, where an actual modeling session takes place. Your alternative is to improvise. A professional photographer I know does all his interior shooting in his living room, which is where this shot of model/actress Amy Miles was taken. (*Photo by the author.*)

he should show in the frame, because some comedians gesticulate more than others.

Whatever type of subject he was shooting, Schaub also planned how to record his subjects. For the model it was not important, since she would not be doing any speaking; he just used the audio dub control of his VCR to lay in a suitable music soundtrack after the actual shooting was finished. For the actor he used a lavalier mike that would pick up just the actor's voice without any extraneous noises around the camera. For the comedian he decided to use the mike that came with the camera, because he wanted to pick up the sounds of people laughing as they watched the comedian perform.

Since he has started doing his audition tapes, John Schaub has found that he has an extra source of income that comes in handy when his photography business lags.

IMAGINATION

There are any number of ways to make money with your video equipment. All you have to do is take a look around you and see what kinds of opportunities present themselves. With a little bit of imagination, you too can find such opportunities.

Epilogue

The home video enthusiast is caught up in the midst of an electronic revolution. No longer do the established television networks and independent stations control the airwaves. Just recently, the Nielsen demographic survey has shown that the audience viewing network programming has been sharply reduced. While no definitive explanation for this decrease in audience size has been forthcoming, speculation within the television industry is that more and more people are subscribing to cable television services, while many others are purchasing videocassette equipment to use with prerecorded tapes. The home videographer, of course, is making his own television programming when he shoots his own tapes and plays them back on his television set.

I do not think any book that deals with the home video revolution can be complete without a brief mention of the poor quality of broadcast television, a factor that has ushered in the rapid rise of home video products. Broadcast television was once a far more positive influence, especially to people growing up.

TELEVISION HEROES

No two words are more evocative for me than *coonskin cap*. Say them and I instantly imagine Davy Crockett fighting the Mexican soldiers on the Alamo battlements and refusing to go down.

Most of my heroes as a child were characters in television Westerns. When I got home from school, I switched on the TV and watched *The Lone Ranger*. Every Tuesday night at 7:30 I watched the adventures of *Bronco*, *Cheyenne*, or *Sugarfoot*, cowboy heroes who alternated from week to week.

Certainly, the clear-cut good and evil characters in those Westerns and the other fifties series were unrealistic depictions of the way people are. However, my heroes still taught me some valuable lessons by example. The Lone Ranger and Tom Brewster, alias Sugarfoot, gave me a respect and love for the law, while Davy Crockett taught me to stand up for what I believed in, regardless of what others thought.

It seems to me today, though, that the time of the true television hero is past. Contemporary television "heroes" now come in four categories: the nonhero, the illiterate, the villain, and the superhero. All are role models for today's young viewers, just as the fifties Western heroes were role models for my generation.

The nonhero first appeared on the small screen after the commercial success of the Clint Eastwood film, *Dirty Harry*. Usually a cop or private investigator, the television nonhero is just as violent, just as unsympathetic, and just as emotionally empty as the "perpetrators" he chases. The archetypal nonhero uses a weapon declared too lethal to be legal in many states (*Starsky and Hutch*), lives in a dirty, one-room walk-up (*Baretta*), and sucks lollipops while violating civil liberties (*Kojak*).

Many of TV's role models for young people today are "greasers" and "sweathogs." This illiterate role model first appeared on television in the person of the Fonz, the hero of *Happy Days*, a series whose original inspiration was the highly successful film *American Graffiti*. Unfortunately, the Fonz can hardly read and, before the 1982–1983 season, never had a monogamous relationship with a woman.

Happy Days spun off *Laverne and Shirley*, the bargain-basement Lucy and Ethel duo of the seventies. The girls actually live in a basement and work in a brewery. That's something to aspire to?

The illiterate role model descended the evolutionary scale even further with *The Dukes of Hazzard*. The Duke boys have never been seen to count beyond ten or read a book. They speak with an accent I have never heard in any southern state, drive their cars recklessly and indiscriminately, and then flash their pearly white teeth after the mayhem is over. I would hate to be the state driving inspector who gives the road test to the kids growing up on this show.

On the prime-time soap operas, the villain has attained the rank of hero. Alexis Carrington, *Dynasty's* femme fatale, recently helped to poison her in-law in order to make money. The "hero" of *Dallas*, the highest rated show on TV, is a sadist named J. R. who takes delight in ruining people, even murdering them when it is financially expedient.

Dallas's Friday night time slot makes the show particularly accessible to kids, who are allowed to stay up late and watch it because there is no school on Saturday. And what kind of behavior can a child learn from J. R.? That you can ruin friend and family alike—and get away with it.

Perhaps as a response to the negative role models on television, the superhero has emerged as "a force for good." Although sometimes they are a bit too juvenile or violent, superheroes such as the Bionic Woman, the Incredible Hulk, and the Greatest American Hero at least provide role-model alternatives for young viewers.

A recent episode of *Hero* finds Ralph, the schoolteacher with the superpower suit, in a quandary. While using his superpowers, he has accidentally endangered an innocent person, so he decides to hang up the superpower suit for good. However, he is not certain he is making the right decision; he still feels a responsibility to help society.

Troubled, he sees in the newspaper that John Hart, who played the Lone Ranger on television from 1952 to 1954, is going to make a personal appearance at a shopping center. Ralph takes his son, ostensibly to introduce him to a real hero. Yet, it is Ralph who watches in awe as the Lone Ranger sits atop Silver, the horse's front legs rearing high in the air.

The next day, Hart, arriving early for another appearance, finds Ralph waiting to ask his advice. Hart tells him that if the Lone Ranger were around today, he would stay to do what was right—to serve justice regardless of personal fears. Ralph decides he can do more good using the suit than not and resumes his crime-fighting career.

At the end of the episode, Ralph tells his class of problem students that the values the Lone Ranger stood for are still viable in today's society. The students laugh and ignore him.

Ralph and I are left with the masked man's legacy: a belief in simple human decency. Most of today's broadcast television "heroes" leave younger viewers with a legacy of selfishness, illiteracy, and hatred.

With the poor state of broadcast television, is it any wonder that the home video industry is thriving? Certainly, there is room for broadcast television and home video to exist side by side. However, until broadcast television improves, it is home video that will thrive and broadcast television that will suffer.